S0-BUV-380

GUSHER

The Search for Oil in America

The saga of oil exploration in the United States is filled with fascinating stories of colorful individuals, vast fortunes made and lost, spectacular wheeling and dealing, shady political intrigue, far-sighted vision and startling strokes of luck. Edwin Drake struck oil in Pennsylvania in 1859 and began the first oil boom. As the demand for oil grew, the search spread to every corner of the land and even under the sea, and the giant U. S. oil industry developed. The search for oil forms a vital strand in the history of our nation's growth.

Other Books by Bob and Jan Young

Fiction

ACROSS THE TRACKS
GOOD-BYE, AMIGOS
SUNDAY DREAMER

Non-Fiction

54-40 OR FIGHT!: The Story of the Oregon Territory
FORGED IN SILVER: The Story of the Comstock Lode
THE 49'ERS: The Story of the California Gold Rush
FRONTIER SCIENTIST: Clarence King
GUSHER: The Search for Oil in America
THE LAST EMPEROR: The Story of Mexico's Fight for Freedom
OLD ROUGH AND READY: Zachary Taylor
PIKES PEAK OR BUST: The Story of the Colorado Settlement
PLANT DETECTIVE: David Douglas
SEVEN FACES WEST

GUSHER

The Search for Oil in America

Bob and Jan Young

Illustrated with Photographs

JULIAN MESSNER  NEW YORK

Published by
Julian Messner, a division of Simon & Schuster, Inc.
1 West 39 Street, New York, N.Y. 10018.

For
Dud & Maxine Pratt

Printed in the United States of America

ISBN 0-671-32373-3 Cloth Trade
0-671-32374-1 MCE

Library of Congress Catalog Card No. 73-139085

Acknowledgments

The authors wish to express their profound thanks to the many oil companies which generously and graciously provided us with invaluable material, along with the assistance extended by the industry's informational and educational associations.

Our thanks, once more, to the Whittier Public Library and especially Mrs. Jane Kerker, whose indefatigable searches and other assistance made our research much easier.

Contents

1

Drake's Folly

Edwin Laurentine Drake silently watched the walking beam, which looked much like a grasshopper doing pushups, drive a steel bit into solid rock. His one employee, Uncle Billy Smith, diligently watched the moving parts punching a hole in the seemingly futile search for oil.

Repeated difficulty had plagued Drake, who had used the fictitious title of "Colonel" since arriving in Titusville, Pennsylvania more than one year before, on May 15, 1858.

Colonel Drake had enthusiastically accepted an offer to drill America's first oil well, believing it would at least be an improvement over what had been less than a successful life for him. The best job Drake ever held was a railroad conductor, a job which paid about $75 a month. Drake had been offered $1,000 a year to complete the drilling job, even though he knew less of the skills needed than a mule knows of holidays. But under the auspices of the Seneca Oil Company, Drake was willing to try.

With his family, Drake moved into the American House in Titusville and began to examine the oil seepages around nearby Oil Creek. He first tried to skim the oil—sometimes with blankets, which were later wrung out, or with boards, collecting the oil in slack water. But when these methods returned only marginal amounts, Drake dug ditches and tried to divert the oil which seeped along the creek into pools, where the oil could be easily gathered. That wasn't worthwhile either, but by this time Drake had become zealous to resolve the recovery of oil. A well was needed, and Drake began a search for the proper man to drill the hole. Many were qualified, because men often drilled salt wells, which were common in the area. But few of the drillers were reliable enough to complete a given task.

By this time, the Colonel had been given the derisive nickname "Crazy Drake." After nearly a year of discouraging work building the pump house, installing a six horsepower engine and deciding upon the best possible site, Drake finally managed to obtain the services of William A. Smith, usually called "Uncle Billy," and his son Sam for wages of $2.50 per day.

Smith, a short, muscular, laconic blacksmith, had drilled numerous brine wells and was considered a steady, reliable worker. With Drake's assurance that he would be compensated for the lumber, pipe and drilling tools, all worth $76.20, Smith and his son moved to a site which became the nucleus for a multibillion-dollar industry. Once at the selected location on Oil Creek, they built a slab-sided derrick. It was a pyramid of boards, twelve feet square at the base and three feet at the apex. It was August before the structures, engines and tools had been installed. Drake inspected it, seeing the derrick as either a monument to foresight and genius or sheer idiocy and folly. Only time would tell.

With the preliminaries complete, Drake gave the signal which started the thump and thunder of the cable tools. Crushing their way underground, they reverberated within the derrick. Rotary drills were unknown at that time, and Drake was using the percussion method. A walking beam was attached to the engine and set into motion by ropes run through pulleys. These were hung in the derrick, from which chisel-like bits were raised and lowered into the earth, crushing the rock as they probed deeper. The resultant debris was raised from the hole by a bailer.

Smith pushed the tools urgently through the overburden of earth and toward bed rock. Trouble developed immediately. The semisolid earth continued to cave in, and water filled the hole.

"It is like drilling through sawdust," Smith complained to Drake. "I have drilled many wells in my time, but I have never run into anything like this. It is bad."

Drake agreed and suggested that Smith close down operations for the day. The well would be a fiasco unless they could devise a method to prevent water entering the drilling area. Historical accounts differ, but between them they found a solution. A pipe, six inches in diameter, would be driven through the earth to bed rock, then drilling continued within this casing pipe. This would prevent cave-ins or the intrusion of water hindering the work. Though Colonel Drake has been given credit for this device, Smith later claimed that he had done the same thing in driving water and brine wells at Salina, Kansas.

No matter who originated the idea, it worked. On August 15, the pipe was inserted and driven down thirty-nine feet to the bed rock, using the force of the small engine and the rhythmic raising and dropping of the bits, which now continued without further interference from cave-ins or water.

Even with the force of the engine, progress was slow. It was late in August before Drake's well had been pushed through thirty more feet of rock. Measurements indicated the bit stood at sixty-nine feet.

It was near quitting time, Saturday, August 27, 1859. Colonel Drake, in his black frock coat and stovepipe hat, was a homely, bearded and lanky man. His face showed concern. He had been hired to "raise and dispose of oil," and so far had failed. The company was nearly out of funds and patience with the project.

As he gloomily considered the future, Drake didn't know that a meeting was going on in New Haven, Connecticut, among Drake's backers. As Drake assumed, these men were conservative bankers and professional men, none noted for their interest in taking chances or losses. They had become disenchanted with "Drake's Folly," and a letter was being composed to cease all work and abandon the project. A money order was enclosed with the letter in an amount sufficient to pay any remaining wages or other debts.

"Drake is fooling away his time and our money," the stockholders complained. They had spent $2,490 for drilling the well, and that was considered enough to lose in such a chancy venture as oil. For the past few weeks, James M. Townsend, a flint-eyed banker, had advanced all of the operating expenses, backing with his own money his faith in the well's ultimate success. But now, Townsend, too, had reached the limit of his finances. No more dreams. No more drilling. The letter was sent to Drake at Titusville.

Neither Drake nor Smith knew that they were both virtually unemployed as drilling was continued a few minutes after the usual quitting time. Smith saw the drilling tools suddenly drop a few extra inches, then standing at 69½ feet. Smith stopped the engine and peered into the drill pipe. Just below the platform through which the drilling tools worked, Smith thought he saw the greenish glint of oil. Excitedly, he yelled for Colonel Drake to come and see for himself.

When Drake ambled up, Smith was preparing a bailer to bring up a sample of oil from the six-inch pipe. It was composed of fine rags, which he rammed down the hole. It was withdrawn covered with greenish oil. Then a fine spray of oil foamed up from the pipe itself. Drake and Smith rubbed it on their hands, smelled and tasted it. Then both men smiled.

"Oil! Really oil," Drake murmured.

Once the excitement of the momentous initial success had been absorbed, both Drake and Smith were matter of fact about the situation. A hand pump was attached to a twenty-foot piece of pipe, which was the first pumping of an oil well in the United States. It was also the first well ever to be drilled in a search for oil.

No one thought to measure the initial issue of the Drake Well, but it later continued to produce about 35 barrels a day. While that was a microscopic trickle in comparison with later discoveries, the Drake Well marked the founding of a giant industry.

"That's your fortune coming out," Uncle Billy Smith smiled as the flow began. Drake, like many other pioneer discoverers, didn't profit much for his work. Inexplicably, he stood aside and watched hundreds of others try to duplicate his success by poking holes in the banks along Oil Creek. Drake didn't become excited when the frenzied stampede to Oil Creek began, a rush which matched the commotion made during the California gold rush days. He continued to pump a few barrels a day, selling as best he could. Because of his success, Drake could have had almost his choice among the leases or shared with any of the men who arrived to search for oil, but he was content to fish and fritter his time away playing cards in the various Titusville saloons. While his original backers seized leases which made all of them extremely wealthy, Drake accepted a position as justice of the peace, ironically formalizing many of the lease agreements in which he should have shared.

Drake finally sold out what assets he had and moved to New York to associate himself with a brokerage firm which specialized in oil properties. It was only a little more than a year before Drake was broke again and without a job. The return of a nervous ailment affected his spine and made him a cripple, barely able to drag himself about.

An old Titusville friend heard of his plight and arranged a public meeting in Oil Creek to consider what could be done to help him. More than $4,000 was pledged immediately, and he was given a home in Bethlehem, Pennsylvania. The state legislature finally recognized the contribution Drake had made to the state and to history. Drake was voted an annual annuity of $1,500, which was continued for his wife when Drake died on November 8, 1880.

About ten years later, Edwin L. Drake was reburied near Titusville, not far from the $100,000 monument which was erected in his honor.

Such was a fragment in the life of a man who helped to light the lamps and to lubricate the axis on which the world turns.

2

Then Came Oil

When Colonel Drake and Smith brought in their historic well at 69½ feet during 1859, they also reached back in time about six hundred million years. It is believed that some oil formations were created that long ago.

Scientists as modern as Dmitri Mendeleev, the noted Russian chemist who devised the Table of Periodic Elements, insisted that oil was formed by the action of sea-water vapors on deposits of carbon—the inorganic theory of oil formation. But his and similar theories have now been discredited, and most scientists believe that oil is entirely of organic nature, though some aspects of the formation are not entirely understood.

The creation of oil is the result of a struggle between titanic forces and curious mutations. The earth appears to be relatively stable, but it is always in flux: contracting, contorting and shifting in both land and oceanic areas. Millions of years ago, seas invaded and washed over the mountains and remained there until the oceans were pushed back by the new growth and upthrust of the mountain masses. Once the mountains emerged from the seas, they in turn became subject to the attrition of nature. Rains, snow, heat, cold and even the persistence of plant life pulverized and eventually leveled the mountains until the seas washed over them once more. The cyclical process has been going on since the beginning of time.

With each succession of seas and mountains, remains of the

previous era were left behind: skeletons of fish, birds and dinosaurs, along with trees and other plant life.

But essentially it was the diatoms and the plankton, the free-floating plants and animals which teem in the oceans in countless trillions, which were responsible for oil. As these died, their tiny remains settled to the bottom in a veritable marine blizzard of sedimentary deposits. These, along with other organic debris which is constantly washed into the seas, decomposed, possibly with the aid of anaerobes, which are bacteria that do not require oxygen to live. As layer after layer sifted down, enormous pressures and head developed, and silts, muds and sands transformed into sedimentary rocks, usually in thin strata. Of this mixture some became limestone, some shale and some sandstone; some of this formed a hard cap over extensive deposits of organic matter.

Microscopic examination of the limestones and sandstones sometimes reveal rocks which are so loosely formed that there is a great deal of space between each particle, much like marbles in a bag. Into these areas sea water seeps, becoming greatly compressed as the weight and pressures from above increase. Thus, after the oil is formed from decomposed organic matters of hydrogen and carbon elements, both oil and water are contained within the rock layers, similar to the way a sponge retains liquids within its pores. The presence of these water-filled sand- and limestones are vital to the accumulation of the worth-while deposits of oil.

As such, there are no "pools" of oil underground. The oil is contained in porous layers or is trapped between nonporous layers.

Oil, being of low specific gravity (that is, lighter than water), tends to rise to the surface of water. Thus, the minute droplets of oil are able to migrate through the adjoining loose-grained stones. This movement may be no more than an inch a year, often less. But there is an inexorable movement of all reserves in the world—toward the surface. This maddeningly slow movement continues until the oil encounters an impervious layer of rock, and there the loose-grained rocks become saturated with oil, which replaces the water. It is in these areas that the most successful wells are drilled.

"Gushers," which are explosively powerful artesian wells, occur when gas pressures, also trapped in the rocks, force the oil to the surface with enormous energy. As a result of gradual distillation of the oil through changes from constant temperatures and pressures, a "natural gas" forms. It, too, is a mixture of hydrocarbons, the basis of all petroleum products. Sometimes wells are drilled and found to contain nothing but natural gas. Many of these so-called gassers contain no oil, but they are extremely valuable for the gas products they issue.

In 1837, an imaginative farmer named Daniel Foster, who lived near Findlay, Ohio, became the first known American to control and use natural gas in a home.

Foster needed more domestic water for his home, and so he began to dig a well. He went down several feet and found no water, but decided to give it one more day's work. When he probed a few feet farther, Foster was almost overcome with foul-smelling gas. He managed to scramble out of the well before he was overwhelmed.

Foster, a thrifty man, tried to think of some way to use the escaping underground gas. Finally he hit upon a plan. Punching a hole in the bottom of a large sugar kettle, Foster capped the well by upturning the pot over the hole, then sealing the edges with clay. The well continued to hiss gas through the aperture. With a helper, Foster hollowed out several small logs to form crude pipes. One was sealed into place on the kettle's hole. Others were attached to the first, the joints wrapped with canvas and smeared with clay. The final length, which was inserted into the cook stove, was an old rifle barrel in which Foster had drilled several holes. Once everything was in place, Foster struck a spark and said to his wife: "Woman, get to cooking. The stove is lit."

Like natural gas, in some unusual cases oil reaches the surface of the earth and exudes in oil springs, seepages and the like. Usually such oil is of poor quality because of the changes brought about by evaporation, oxidation and similar factors. Such surface oil is but a minute fraction of the total available petroleum. Most of the oil is far below the surface of the earth, contained in "reservoir rocks."

Oil geologists have devised a table of time from which petroleum could be taken, though the date fixing is somewhat arbitrary because of the millions of years involved. From rocks of the Tertiary/Cenzoic era 39.2 per cent of the oil is recovered. This was formed about one million years ago.

A smaller amount of petroleum, 14.3 per cent has been found in deposits from the Mesozoic era, which dates back to the time when enormous downfolds, or geosynclines, formed and the earth was washed with the seas again.

Most important was the Paleozoic era, in whose strata 46.5 per cent of the oil is located. Plant life and reptiles flourished, forming the organic material from which oil is derived.

Some of these figures are educated guesses by oil experts, but they are as accurate as can be obtained. There does not seem to be a strong correlation between the type or quality of crude oil and the age of the strata from which it is taken.

While the formation of oil is dated by oblique methods, no firm date can be assessed for the first discovery of oil. Drake's well was, of course, the first drilled in a determined search for oil, but there is evidence that oil was known and used for at least five thousand years before 1859. It was found on the surface in various forms.

More than four thousand years ago, the Sumerians, Assyrians and Babylonians were using asphaltic bitumen, which is a petroleum in a viscous, almost solid state. It was used in numerous ways. Crude oils were stirred in with ground paints, a combination which gave the mixture a firmer, adhesive quality. These may have been the first "oil paintings." Ships, baskets, wicker mats and other materials were treated with this petroleum to waterproof and preserve them.

A grimmer use was made of the boiling asphalt when it was launched from catapults toward enemies or poured over the heads of condemned criminals.

For thousands of years petroleum products have been used to ease the pain of rheumatic joints, taken as a purgative and used in the treatment of external sores. When mixed with beer, petroleum was claimed to be a specific cure for inflamed eyes as well as other ailments, though no mention was made whether this ghastly potion was consumed or merely rubbed on affected areas.

Because of its random occurrence, structure and origin, petroleum acquired the characteristics of magical properties. Where ignorance is supreme, mysticism flourishes. But the ancients found practical ways of using petroleum. It was heated by dropping small stones into it, then applied as paving for roads. Egyptians stuck some of their pyramids together with various mixtures of petroleum. They also embalmed their dead with fluid oils and filled body cavities with asphalt. The remarkable preservation of mummies is an example of this practice.

Because petroleum occurs in various colors, places and degrees of fluidity, the substance has acquired many names: asphaltum, slime, pitch, matha, nephar or naphtha. The name "petroleum" is most apt and is of Latin origin: "petra" meaning rock and "oleum" meaning oil.

"It is a kind of slimy mud which will burn clear," one ancient historian wrote. "When it meets with anything solid or hard, it sticks to it. . . . If it is touched, it follows them that flee from it. By this means, the townsmen defended their walls and the enemy fried and burned within their armor. . . ."

Intersticed throughout the leaves of history is mention of petroleum. In 431 B.C., Princess Crusa was said to have been destroyed by Medea when she was given a robe saturated with petroleum and then set afire. "Whenever she shook her locks, the flames started up twice as high. . . . She sank to the floor, past recognition save to a parent's eye."

The Book of Genesis, 6:14, giving instructions to Noah on the building of the ark, God tells him to make it of gopher wood and "pitch it within and without with pitch."

The Book of Exodus, 2:3, describes the way in which the infant Moses was concealed by his mother. ". . . She took for him an ark of bulrushes and daubed it with slime and pitch. . . ."

Despite this widespread recognition of petroleum, scant use was made of it because of failure to recognize its unexcelled value as a lubricant, fuel or illuminant. Most of these modest needs were then supplied by plant, animal or fish oils.

Wider knowledge of petroleum was spread by far-ranging sailors. The wooden ships of the time were subject to the ravages of many marine borers and treatment with petroleum reduced the hazard from such attackers. Storms often wrenched planks apart, which required new caulking and treatment by petroleum. Though pitch from trees served admirably for these purposes, it was not always available and certainly not always satisfactory because of its tendency to melt in tropical areas. Loss of pitch from the ship's seams meant disaster. The discovery of a source of petroleum by Spanish sailors in 1526 was therefore of the greatest importance:

"... There is in the Islands of Cuba certaine Fountaines at the Sea Side that doth cast from them a kinde of blacke Pitch of a strong smelle, whiche the Indians doe use, in their cold informitives, oure people doe use it there to pitch their shippes withall, for it is well nere like unto Tarre," said one account, written by a ship's surgeon.

It was more than a half century later, on March 22, 1595, that Sir Walter Raleigh wrote in his journal, regarding a "Pitche Lake" off the South American coast:

"... At this point called Tierra de Brea or Pitche there is an abundance of stone Pitche that all of the ships of the world therewith laden from thence, and we made a trial of trimming our shippes to be most excellent goode, and it melteth with the Sunne as the pitche of Norway and therefore shippes trading with the south parts profitable."

As it still does today, petroleum played an important part in international affairs, even in the earliest days of commerce. France took the lead in staking out a share of the New World, seeking whatever treasure might be found there. France moved in force at a time when England was staggering from internal and external wars and beset with trade and other difficulties.

French missionaries were dispatched to the Great Lakes area of Canada, and they moved southward from there into the Louisiana Territory. Two priests, Dollier and Gainee, in 1670 reported a "fountaine de bitume" near what later became Cuba, New York. The reports of these Sulpician priests were later confirmed by Father Joseph Daillon, who wrote of "a good kind of oil which the Indians call Antonontons."

Other seepages and petroleum deposits were described by the wilderness wanderers as "a thick oily stagnant water, which would burn like brandy."

Many years passed before an American cartographer, Lewis Evans, marked his map with known oil sites. Evans pinpointed seepages at a place now called "Oil City" and marked a "petroleum" location there.

Gradually the control of vast areas which later became a part of the United States shifted from France to Great Britain, then to the Americans, and approaches to oil exploitation became more aggressive and pragmatic. By that time, the empirical science of chemistry had emerged from the black magic of alchemy. Quite naturally, the American genius Ben Franklin became interested in uses for oil. He experimented with the use of oil to quiet the ocean surf in an effort to make harbors and landings safer. His finds were later consolidated into a monograph, which was read before the Royal Society in London on June 2, 1774, and later published among the Society's *Philosophical Transactions.*

Though Franklin hedged his experiments by using whale oil, rather than any crude or refined petroleum, his foresight for such a use anticipated other thinkers by more than a century. The British, during 1880s, made intensive investigations along the line which Franklin had devised. They laid down pipelines in Aberdeen harbor which were designed to release oil and calm particularly turbulent conditions.

In 1767, a Moravian missionary, David Leisberger, described three different types of petroleum in the Allegheny River regions. He told of seeing Seneca Indians take woven baskets to where oil seepages flowed into the river. The Senecas stirred the backwaters with long sticks, then they waited until the oil floated to the surface again and accumulated in a viscous mass. The Indians then dipped up the oil-water mixture in their baskets. This concoction was boiled by dropping hot stones into the containers. With the water boiled off, the remaining pure oil was used for toothache, swellings, rheumatism and sprains. "It was also used some for lighting their torches," the missionaries observed.

Though scant determined attention was paid to oil locations during the turbulent days when the thirteen colonies were fighting for freedom, a significant letter was written by General Benjamin Lincoln to the Reverend Joseph Willard, president of University of Cambridge. In 1783, Lincoln recalled, his troops had been bivouacked at an oil spring, which was later determined to be along Oil Creek, where Drake brought in his well. There, Lincoln said, his men collected "Barbadoes Tar," often as much as seven gallons a day, which they used to treat ailments. "It gave them great relief and freed them immediately from many complaints," he said.

As the area near Oil Creek was gradually inhabited, more interest was generated about the use of the seeping oil. ". . . A spring tributary much celebrated for its bitumen, resembling the Barbadoes Tar, was known by the local name of Seneca Oil," Joseph Scott commented in reference to the oil springs in 1795, when he compiled his "Gazeteer of the United States." "One man could gather several gallons a day of this sovereign remedy. . . ."

Despite the increasing number of reports of petroleum finds, exploitation remained somnolent because there were more than seven hundred stately whaling ships scouring the seas for the great sperm whales, from which "ile" could be extracted. Though the value of the "sovereign remedy" was appreciated, few were excited about discovering more so long as men went down to the sea in ships. It was safer than men going down grubbing the earth. Most American homes were lighted with whale oil lamps or tallow candles.

Scientists, continually curious about any subject which implied new vistas and progress, were anixous to explore the petroleum prospects in view of the indications that other lubricants and illuminants might become depleted. Animal fats, cottonseed, vegetables and fish derivatives were tested and refined, but all of these were expensive, and some combinations were extremely dangerous. With the added specter of depletion, industrial chemists looked to the extraction and refining of oils from coal tars and asphalts. For a start, the immense asphalt lake on Trinidad seemed to offer an unlimited source of raw materials.

By 1830, Samuel L. Downer, of Boston, had perfected a refining process which efficiently extracted the illuminant and lubricating elements from various petroleum substances, and he was able to market his products at a profit. About twenty years later, Downer acquired a process from Luther Atwood to produce "coup oil," a lubricant which Atwood derived from coal tar. The coup oil was extensively used by railroads and cotton mills until better materials were properly refined and available in great quantities.

There were others working in the same field, almost unknown to each other. When they were aware of others, they derided the work being done.

Abraham Gessner, a chemist, inventor and later a doctor, continued his research in trying to improve methods for coal oil. In 1846, Gessner, a Canadian, developed a process of securing illuminating oil from cannel coal, a low-yield petroleum amenable to refining through heating and evaporation. Gessner patented his product under the name of "kerosene," a name which has now become generic for all illuminating oils extracted from both coal and petroleum substances. The name is a contraction of Greek words meaning "wax" and "oil."

Before the American industry began to flourish, another industrial chemist, James Young, a Scotsman, started manufacturing refined oil taken from oil springs which suddenly appeared near some coal mines at Derbyshire, England. The oil spring was shallow and quickly exhausted. Young then resumed distillation of a paraffin-based oil, a method which he patented in 1850. His refinery was enormously successful. Young admitted to a profit of 300,000 English pounds in one year. (In those days, that sum was valued at nearly $1.5 million dollars.) In addition to lubricating oils, Young manufactured a "Brown Bathgate Naphtha," which was used to waterproof garments because the naphtha was a specific solvent for rubber and was used to treat clothing. But when he heard of the Drake well and subsequent oil production from the Titusville and Oil Creek area, Young scoffed, in one of the world's greatest understatements: "It [the oil business] is ephemeral and won't last."

3

The Stage Is Set

Nathaniel Carey, a pioneer settler along Oil Creek, called modest attention to the potential of oil when he began hawking a patent medicine. In Pittsburgh about 1790, Carey began selling what was called "Seneca Oil" (after the great Indian Chief), skimmings from Oil Creek carried in five-gallon kegs. The oil was sold to apothecaries, some say for $80 (16 per gallon), which then bottled it in two-ounce containers.

Carey's production was modest, and because he was the town tailor at Franklin, he was willing to take food, tailor supplies or other goods in exchange for his oil. But Carey's success, however modest, attracted competition, and the limited market for Seneca Oil was quickly glutted. Most families would use no more than one bottle a year, and the small amounts skimmed from the Creek were ample to fill the demand. But oil was discovered while drillers searched for another product: salt.

Salt was a vital commodity in pre-refrigeration days, not first known salt well west of the Alleghenies was drilled in only for meats and perishable foods but for seasoning. The 1806 when two brothers, David and Joseph Ruffner, of Virginia, put down a well in Kanawha County of West Virginia.

Their father's farm, situated along the Kanawha River, contained a large surface salt lick, and the Ruffner brothers believed that high-quality salt could be obtained by drilling there and bringing up the underground brine. But the idea came much easier than its fulfillment. It took the young men eighteen months, much ingenuity and more patience to make the well because of quicksand, which made drilling extremely difficult.

Being pioneers and knowing nothing whatever about drilling, the Ruffners had to improvise every move. To offset the difficulties of working with quicksand pouring into their well, they dug to bed rock, seventeen feet below, and then set a hollowed sycamore tree, four feet in diameter, into the hole. This was called a "gum." The arrangement was held in place by a platform from which the muck and mire were bailed, leaving the hollow tree, the "gum," virtually free of water and sand. Then, with a 2½-inch bit, the Ruffners began to bore into the bed rock, using the "spring hole" method, a means which was also called "jigging a well."

Jigging down a well was merely the use of a springy green sapling about forty feet in length and usually a foot in diameter. The butt of the sapling was fastened to a post or into the ground. Another post was erected about ten to fifteen feet from the butt and acted as a fulcrum. The pole was then adjusted to reach several feet beyond where the drilling bit would be attached to the punch the well hole. With the tools in their proper places, the Ruffners applied their total weight to the small end of the springy sapling using a stirrup arrangement. They alternately pulled it down and released it, making a short drilling stroke each time. It was arduous work, but these primitively dug wells were punched several hundred feet in the earth. The Ruffners are generally given credit as being the first to introduce this method west of the Alleghenies. As they punched their brine well deeper and deeper, the Ruffners inserted a pipe. Because there were no pipe screw joints then, the Ruffners invented the "seed bag sealer," a device which prevented water or sand from clogging the drill pipe. The sealer was simply a leather sleeve drawn over the pipe intersections. The sleeve was cut to fit as tightly as possible, and it was filled with dry flax seed, which was soaked with water before being fitted into place. As the flax seed swelled, a tight seal was made at each joint.

Though the task had thoroughly tested their mettle, they finally made a well. At fifty-eight feet, the Ruffners struck a strong flow of brine, which was mixed, or tainted, with unwanted oil. Once more the Ruffners had to improvise to make their well. They diverted the brine into a cistern and allowed the oil, which rose to the surface, to overflow into the Kanawha River. The Ruffners didn't bother to retrieve the oil for commercial sale. There was too much available already. Salt was the important product from the well. Other such salt wells in Kentucky, Ohio, Tennessee and Pennsylvania were plagued with the oil content in their brine wells.

If these pioneer oil men had only known.

One Kentucky driller dramatically demonstrated how little oil was wanted, valued or understood. Starting the "American" well in Cumberland County, the brine well driller dramatically announced: "I'll strike salt water or I'll strike hell."

He too should have known better. Not long before he began work, a Wayne County, Kentucky, driller made a well which was heavily contaminated with an "odd, ugly grease." The flow was accidentally ignited after it had overflowed into the Cumberland River. The ensuing fire, which was estimated at fifty miles in length, could be neither controlled nor extinguished until the oil ceased to flow. This well, considered to be the world's first known gusher, shot a column of oil one hundred feet into the air.

Despite this fiery example before him, the American well driller defied nature in making his statement. Worse, he had a forge near where the well was drilled, and when he punched a hole 180 feet deep, a heavy flow of oil surged into the drill pipe and began to flow. A spark from the boilers ignited the oil, and the driller yelled: "I've struck hell. I've struck hell. God have mercy." Then he fled the scene.

That fire was eventually extinguished when the oil ceased to flow. When it resumed artesian issue, it was then bottled and sold as American Medicinal Oil. It competed successfully with the pioneer Seneca Oil because the two contained identical properties.

Though the sovereign value of Seneca Oil was accepted more or less on faith, there was little really known about the real properties of oil. Though his son, also a Yale chemist, was considerably more perceptive and enthusiastic about the value of oil, Professor Benjamin Silliman, Sr., of Yale, considered one of the foremost chemists of his time, examined the oil spring near Cuba, New York, in 1833. He was probably the first to refine it by heating and then straining it through a porous cloth. He observed that the end product was used as an external treatment for boils and other sores, particularly upon horses, not an attractive market for large producers.

Many years later, his son, Professor Benjamin Silliman, would issue what has been described as a landmark report on the value of oil and its possible derivatives, such as the extraction of paraffin and illuminants and candle manufacture from various substances. But even before either of the Sillimans became involved in the burgeoning petroleum industry, a Marietta, Ohio, physician, Dr. Samuel P. Hildreth, said in 1833 that oil had great possibilities as a lubricant. He wrote that skimmed oil was used in lamps where the substance was abundant, though its advantages as an illuminant limited because of its usually offensive odor. Hildreth observed that the substance could probably be improved by filtering it through something like charcoal. Unfortunately, his theories weren't accepted for a long time, else the oil industry would have had its inception before the Drake Well.

It wasn't until 1845 that Lewis Petersen, Sr., of Tarentum, experimented with a small amount of petroleum mixed with sperm whale oil and found it to be an excellent lubricant for the spindles at the Hope Cotton Factory in Baltimore. Because of the savings attendant to the use of this oil, Petersen was sworn to secrecy, and he supplied the factory with two barrels a week for ten years before the secret became generally known.

In time, the use of oil as a lubricant was vastly more important than its initial recognition as a patent medicine, but American Medicinal Oil had a significant impact on the future industry. One bottle of the American Oil displayed in a druggist's window was noticed by Samuel M. Kier, a Pittsburgh businessman. He had been born in Pennsylvania in 1813 and attended common schools there. When he was twenty-one, Kier set out on his own. He worked at a variety of jobs but returned home when his father bought a large farm near Tarentum in 1847. Young Kier decided to put down two brine wells, but both flowed oil instead of the desired salt water. Kier used the crude oil in his furnaces, which were used to evaporate off the water from which the salt was obtained. Kier got the idea that the oil might have other uses too. He had an industrial chemist analyze samples. The chemist suggested that the substance might make an illuminant, if it was refined and some of the impurities were removed.

Kier obtained a small still, an apparatus which was nothing more than a one-barrel, cast-iron pot in which the oil would be heated. To gather the wine-colored distillate, a circular tubing (called a worm) was inserted. The refined liquid was called "carbon oil." Unfortunately, Kier found that the carbon oil would not burn properly in the lamps then in use. He tried to design a lamp which would accommodate the carbon oil, but even his best lamps still gave off an offensive odor. Kier gave up that aspect, and the problem wasn't solved in America for a decade. For a time he allowed the oil to rise to the surface of the brine and overflow into the Pennsylvania Canal, a practice which was prohibited after several disastrous fires occurred from Kier's and other wells' overflow.

About a year after Kier moved to his father's farm, his wife became ill. An attending physician prescribed regular doses of American Medicinal Oil. With the first spoonful, Kier immediately recognized that this nostrum had the same vile odor, taste and ignition properties as the crude oil flowing from his wells. Kier saw that he had been wasting a fortune. Next day, he began funneling his oil into barrels. Then he began marketing small amounts under the label of "Kier's Petroleum or Rock Oil." Accounts vary, but a half-pint bottle sold from fifty cents to one dollar each.

Kier had the soul of a salesman. The bottles were gaudily labeled and were sold from a fleet of fifty wagons, which were also garishly decorated with illustrations of the Good Samaritan ministering to the ill within the shade of a palm tree. Kier hired teamster-salesmen to rove through outlying areas to hawk the oil. After some modest entertainment, followed by a high-pressure sales pitch, Kier's salesman sold the bottles to the gathered crowds. When the country was too sparsely settled to attract a medicine show, the oil was sold from farm to farm. Kier managed city sales, placing the bottles on druggist's shelves, windows or wherever they could be profitably displayed. Their labels said:

KIER'S
Petroleum or Rock Oil
Celebrated for Its Wonderful Curative Powers

A Natural Remedy
Procured from a well in Allegheny County, Pa. Four Hundred
Feet Below the Earth's Surface

Put Up and Sold By
SAMUEL M. KIER
No. 363 Liberty Street
Pittsburgh, Pa.

Usually accompanying the displays were testimonials attesting to the worth of the oil. One display simulated a used Pennsylvania $400 bank note, indicating that it had been drawn on the Bank of the Allegheny River.

Despite the fact that Kier's Oil was guaranteed to cure colds, coughs, cholera, morbus, burns, bruises, old sores, ulcers, bronchitis, asthma and a number of other debilities, including blindness (!), it never sold briskly. It was not as successful as the old standard, Seneca Oil.

Lubricants and illuminating liquids were vital to the burgeoning growth of the United States. Ships, trains and machinery could work more efficiently and safely with both of these substances. All America wanted a safe, cheap source of light, one which would replace the dangerous and expensive camphine, a mixture of alcohol, ether and turpentine.

Little was apparently known of James Young's success in refining, though the Downer Company of Boston was producing Coup Oil in large quantities. Two hundred thousand gallons accumulated before a new design in lamps absorbed the backlog, making a $100,000 profit for the company.

With profits of that size as the potential prize, many were willing to take the risk, and fifty distillers opened as quickly as suitable materials could be assembled. With the expansion of the illuminant market, the future looked brighter. Then an unexpected obstacle appeared. Surface seepages of oil were waning, and the budding industry was threatened with a lack of adequate supplies of oil.

Kier, who had shown the way and was considered as America's first commercial refiner, bought ten acres in the Oil Creek area. He tried to buy more acreage from the lumber dealers Brewer, Watson & Co. Shrewdly, they refused to sell, since the commercial use of skimmed oil had been demonstrated by Kier and others. Besides, the employees in the sawmill used the oil to treat their saws and sometimes used the skimmed oil for fuel. The company did sell lease rights for five years to J. D. Angier on acreage owned in the Cherrytree Township. Angier dug trenches around the oil springs located on the leasehold. Oil and water was gathered in a basin and the oil skimmed off. But when Angier's return was only six gallons a day, he forfeited the lease rights.

However, Dr. Francis Brewer, a part owner in the lumber company along with his father and others, wasn't entirely discouraged. He took a small bottle of the oil skimmed from their oil springs to his uncle and former teacher, Dr. Dixi Crosby at Dartmouth College, perhaps for an examination, perhaps as a gift. At least, nothing seems to have come of the visit at the time.

Not long afterward, Henry Bissell, also a Dartmouth graduate and at the time a lawyer, visited Crosby. Bissell was excited when the Professor suggested the oil would make a superior illuminating oil. With his law partner, Jonathan Eveleth, Bissell suggested they send the professor's son, A. H. Crosby, to inspect the Oil Creek area, and if his report was favorable, a joint stock company would be formed with Brewer. The report was enthusiastic, but Brewer rejected the arrangement. He sold out his interests in Brewer, Watson & Co., and established himself as a banker in Westfield, New York.

Notwithstanding, Bissell and Eveleth then formed the Pennsylvania Rock Oil Company and bought about 100 acres for $5,000 on December 30, 1854. Most of the stock was issued to the original partners. Luther Atwood was asked for an analysis, and Professor Benjamin Silliman, a Yale graduate like his illustrious father, was hired for $1,200 to examine the oil springs and issue a report. His study, which took five months, was enthusiastic. Silliman said in part:

". . . It appears to me that there is much ground for encouragement in the belief that your company have in their possession a raw material from which, by simple and not expensive process they may manufacture many valuable products. It is worthy of note that my experiments prove that *nearly* the *whole* of the raw products may be manufactured without waste, and this solely by a well-directed process which is in practice one of the most simple of all chemical processes. . . ."

Silliman also offered to make firm suggestions about how the raw materials might best be utilized. It was a magnificent endorsement, and then when the company was reorganized for further financing, Silliman was given two hundred shares of stock and named first president of the company. Others included in the company were Brewer, Watson & Co., Bissell, Eveleth and James M. Townsend, president of the City Savings

Bank in New Haven, Connecticut, who would later play an important part in bringing in the first well.

Because no one thought of boring a well just then, Angier was commissioned to resume work with his network of ditches and basins. When the production proved inadequate to continue work, a squabble erupted between the various stockholding factions. By 1855, work had ceased and the company languished.

As often happens in history-shaping events, the change of course was brought about by sheer accident.

In 1856, Bissell was standing in the shade of a drugstore awning as he waited for a friend to arrive. He scanned the window displays, noting Kier's simulated bank note for $400 drawn on the Bank of the Allegheny River. What excited Bissell was the statement that Kier's Rock Oil had been extracted from a well four hundred feet deep! Bissell was sure that deep wells were the solution to the recovery of oil, and his partner, Eveleth, immediately agreed. Together they were able to convince the stockholders that new stock should be issued, and work resumed, this time boring a well. Townsend took a large portion of the stock, and a drilling firm was hired to bore the hole, but it went broke in the financial panic of 1857.

With the bankrupt drilling contractor went Townsend's money, and he was infuriated, claiming that Bissell had merely used the device to unload his stock. Relations between the stockholders, all of whom could foresee no end to the number of holes into which their money might be poured, were strained. When the title to the land on which they had been working was found to be clouded, Townsend moved to oust both Bissell and Eveleth. Concomitantly, Townsend obtained the services of Edwin Drake, who was to attempt to clear title by obtaining two vital signatures as well as inspect the Oil Creek property while in the Venango County area.

Drake had invested $200 of his own money, though he knew nothing about the value of the stock and perhaps even less about oil, even before receiving the assignment of inspection. Townsend was later accused of hiring Drake because he still held a railroad pass which enabled him to ride to Titusville and back without charge to the company!

Whatever the reason, Drake obtained the needed title signatures and made the best oil survey and report he could. While he was there, Townsend shrewdly addressed mail to Drake with "*Colonel* Edwin L. Drake," a fictitious title to which Drake was never entitled but one he never denied. The move put Drake in a prominent position among the residents of Venango County, a posture which might prove most valuable if Townsend believed that stock promotion was necessary later on. With the land and Drake's title both perfected, Townsend moved into power. The Seneca Oil Company was organized and Drake was appointed general manager, though he owned little more than a token amount of stock.

Colonel Drake became zealous in his new job and rightly earned the sobriquet of "Crazy Drake." He was able to open the so-called age of illumination and spud in the first well of the future mammoth oil industry.

4

Oildorado

Colonel Drake triggered the stampede and then stood aside to let the throngs and fortunes pass him by. Not only did Drake not obtain anything of enduring value, but he later stood accused of conspiring with the devil.

"... God put that oil in the earth to stoke the fires of Hell," one pious, irate Titusville preacher scolded Drake. "Surely you wouldn't thwart God and let the sinners go unpunished?"

Drake's reply isn't recorded, but the preacher's imprecations certainly didn't deter the men and women who swarmed into the Oil Creek area shortly after the Drake Well came in.

Jonathan Watson, a partner in Brewer, Watson & Co., was one of the first men on the ground. He arrived before dawn on August 29, 1859, two days after the oil foamed up in the drill pipe. Watson immediately obtained leases on numerous farms where oil springs or seepages were known to exist. On September 1, Watson filed the first lease agreement in the Venango County Courthouse. There, Watson put down several successful wells, one on a site located by a clairvoyant, who was given $2,000 after the well began producing thirty barrels a day.

Watson became one of the first big spenders as a result of the discovery. Money poured into his pockets and bank accounts from the royalties in the Petroleum Center and Oil City wells. Conservative estimates indicate that Watson had about five million dollars on hand within a few months after he began operations. He set aside one million dollars as a working reserve. He had a palatial mansion constructed in Titusville whose grounds were landscaped with exotic plants that cost $50,000. He continued to spend money as if it was going out of style. Unfortunately, within a short time all of his wells had petered out. New ones were dusters, dry holes containing no oil. Soon Watson's million-dollar reserve melted away, and he died a poor man a few years later.

Unlike news of gold strikes which immediately excited everyone, the reports about the Drake and other wells didn't immediately spur a frantic rush. Communications were uncertain and slow. Vague rumors circulated in New York, Pittsburgh and elsewhere, but Townsend's policy of fanatical secrecy staunched the usual flow of information. Townsend, an ultraconservative banker, didn't want his name associated with any gambling or losing proposition. Nevertheless, the *New York Tribune* on September 8, 1859, published a detailed article about "Discovery of a Subterranean Fountain of Oil." Even though the story was widely reprinted, few understood the over-all significance or use of oil. It was almost a year before throngs moved into the Oil Creek area. The first exploitations had been made by the pioneer families and the men who started it, such as Bissell, Kier and Eveleth, all of whom made fortunes—and Townsend, who didn't.

Oil booms differ from gold or other mineral discoveries. It requires considerable capital to drill a well, build the derricks, buy engines and tools, hire drillers and provide for marketing and packaging costs. Gold in the early days could be found and exploited by almost anyone with a pick, shovel and pan. Despite the physical and financial obstacles, literally hundreds of wells were jigged down. Some were spudded in where surface oil seeped. But more were bored by a unique class of oil men, the "wildcatters," who relied more on luck, intuition and other factors than on drilling where oil had already been proved to exist.

One well, owned by Rouse & Meade, produced only five barrels a day. One backer, Bill Barnsdall, insisted that the well be drilled deeper to perhaps produce more oil.

"It is a long way to China from the bottom of that hole," Barnsdall insisted, "but I am bound to bore for tea leaves if I don't hit oil first." His argument prevailed, and the subsequent drilling brought in a well which produced fifty barrels a day.

James Evans, a blacksmith in Franklin, not far from Titusville, was interested in the Drake and other wells. The water well behind his house, situated on French Creek, which ran through the town, often tasted of oil, and Evans decided to deepen his well to determine if oil in quantity lay below. A derrick was built of stripped logs brought in by Evans' youngsters. A few tools, lengths of pipe and bars of iron were obtained on credit, and work was begun. Evans jigged to seventy-two feet when a twenty-barrel well was brought in. Each barrel of oil was worth about $30.

Evans' success spurred another rush. A doctor, hearing of the Evans well, rushed from his house wearing only trousers and, when he reached the site, promptly fainted from the excitement. Courts were adjourned. Bells rang, children cried, dogs barked and at a nearby prayer meeting the evangelist closed the session beseeching a shower of blessings from the Lord. Someone in the congregation said, "Amen, Lord. Twenty-five barrels of blessings."

From this strike, an enduring legend of the oil fields arose. A young man who had been courting Evans' daughter had been out of town when the well was brought in. He knew nothing of it until he called on his sweetheart that night. Miss

Evans came to the door and said rather imperiously: "I don't have to marry you—Dad's struck ile."

Her comment of rejection became a classic and a standard remark when a successful well was drilled. The drillers, tool dressers and roughnecks all shouted: "Dad's struck ile."

Naturally, Evans' well set off a flurry of drilling in the French Creek and Franklin region. There were an estimated two hundred wells kicked down within days after the blacksmith's success. But other wells proved to be short-lived, if they indeed produced anything at all. Within a few months there were more than fifteen producing wells and their output was little more than a trickle, about one hundred and forty barrels a day.

But the excitement and prospect of wealth was contagious. Wells were kicked down from Tidioute to Franklin, with varying results. Most of the town flourished on an ephemeral paper economy.

"Nine out of ten, if not nineteen out of twenty, wells sunk since 1859 have not repaid their first costs, and a portion of the residue wells have done little more than return costs and operating expenses," William Wright wrote in his book *Oil Regions of Pennsylvania.*

One major exception was a lease on the Buchanan farm on Oil Creek. The lease rights were obtained by Henry Rouse and John Mitchell, who began drilling during April of 1860. The land was lumpy and marshy. They chose a low spot to jig in the first well. Slowly the bits probed to three hundred feet, and even the surface prospects looked so promising that the drillers looked for a powerful flow. Still no preparations had been made to barrel the output; nor was any precaution taken to prevent or control fires.

Work went on for a year, and on April 17, 1861, the drillers cut through "a monstrous big vein of oil." A man was sent to tell Rouse, George Dimick (Rouse's cashier) and Mitchell, who were sitting in a nearby shack discussing the fall of Fort Sumter.

Dimick rushed out to obtain barrels, while Rouse and others ran towards the well, which was already spouting a stream of oil higher than the derrick. The gas forcing the oil into the air was hissing so loudly that the men had to shout to be heard. Drillers later estimated the low spots around the site were filled with oil and all of the surrounding ground was saturated with it. Workmen rushed in with shovels to bank the oil pools and save as much as possible.

Work was frantic, and everyone was dripping with oil. In a whoosh, the entire area took fire. The flames raced with the speed of a fiery tornado toward adjoining buildings, then to the spectators and workmen near the gusher. Everyone, everything became a torch.

". . . So numerous were the victims of this fire and so conspicuous, as they rushed out, enveloped in flames, that it would not be an exaggeration to compare them to a rapid succession of shots from an immense roman candle," Dimick observed later.

Rouse was in the center of the fire ring. He sprinted for the edge and possible safety. Even as he ran, Rouse seemed to realize there was scant hope. He extracted his wallet, which contained valuable papers and a large amount of money, and hurled it to the edge of the fire, where it was later recovered intact. He stumbled and then, being unable to right himself, fell. Momentarily, Rouse buried his face in the mud to avoid breathing in fumes and fire. He then regained his footing and resumed running, even though his body was now fully covered with flames. He reached the far edge of the fire. Men grabbed him and carried him to a shanty, his body little more than a skeleton with dangling strips of flesh.

Out of reach of the fire, Rouse lay in the shack writhing in excruciating pain. He remained conscious long enough to compose a will. He refused any religious comfort. "My account is already made up," Rouse said between grimaces. "If I am a debtor, it would be cowardly to ask for credits now." With that said, Rouse died.

The oilfield holocaust raged for three days. Before it was controlled, nineteen men died. Thirteen others were maimed or horribly disfigured. Despite this ghastly precedent before them, leasers continued to drill without any more safety precautions than previously, which meant none at all.

Perhaps the risk was worth the returns. Wells continued to issue vast amounts of oil. One of the best was the Columbia Oil Company lease on the Story farm. In ten years ten million dollars' worth of oil was sold, and during the first years the stockholders were repaid forty-three times for their original investment. One man invested $200 in 1862 and in the next eight years received $107,000 in cash dividends plus another $25,000 in stock shares. That fortunate speculator was Andrew Carnegie, who parlayed his money into one of the greatest fortunes in America. With his fortune Carnegie became a great philanthropist, as the thousands of Carnegie-endowed libraries and schools attest.

Wells jigged down on the Tarr and McElhenny leases brought in major returns. By mid-1862 there were 495 wells flowing, being pumped or being jigged down. Of the three million barrels of oil produced in America during 1862, about 80 per cent of it came from leases in the Oil Creek area. But human greed took a hand in the oil game.

The market for oil was rapidly reaching a saturation point,

but no leaser was willing to control his product or perhaps temporarily cap a fertile well, fearing that the wells would not resume production. This often happened, because haphazard drilling dissipated gas pressures improperly.

As a natural consequence of the overproduction, the price for oil plumetted. Forty-two gallons of oil, one barrel, sold for ten cents, then five cents!

Refining processes improved, with a few high-quality products being extracted from the crude oil. Better lamps and stoves, which could safely use the kerosine distilled from oil, brought about a resurgence of the market. Money again became plentiful, and hundreds or thousands of dollars were exchanged as rapidly as five- and ten-dollar bills had been a few months previously.

Most of the old time residents of the Oil Creek region used their money wisely and well, buying more land and improving what they already had rather than making a showy display of their wealth. But there were others who couldn't stand prosperity, and their legends are still repeated.

One *nouveau riche* sent to Pittsburgh for a case of olives, explaining: "All rich people eat olives." Others bought saddles worked with silver, or boots and dresses. When one Oil Creek housewife was told they had hit it rich, she asked: "Do you suppose we can get a new ax? Mine has plenty of nicks, and chopping wood is hard." Others broke all of their dishes or furniture or took their clothes into the back yard and burned them.

James S. Tarr, on whose farms some fabulously productive wells had been brought in, wisely invested his money and then enrolled his daughter in an expensive finishing school for young ladies. After a brief attendance, Mr. Tarr was informed that his daughter had no aptitude for learning. Tarr jumped into his buggy and headed for the school. The head mistress confirmed the report.

"That's all right. Don't worry," Mr. Tarr said. "I have plenty of money, and I'll buy her the best aptitude possible."

The *ultima Thule* of prodigality was a man named John Washington Steele, who was later best known as "Coal Oil Johnny." Because of his wild conduct and notoriety, a cigar and a brand of soap were named after him, along with a song which spelled out his antics. He also became a favorite target for editorial writers and pastors, who inveighed against his extravagances.

Well they might. In a little more than six months, from the autumn of 1864 to the summer of 1865, Johnny threw away about one million dollars. There are estimates that it was a lesser sum, but it was nevertheless an immense amount.

Coal Oil Johnny was born in Sheakleville, Pennsylvania, in 1843 and was orphaned when he was only a few years old. Johnny and his sister, Permelia, were adopted by the Culbertson McClintocks, a pioneer farm family who owned about a hundred acres of land four miles north of what later became Oil City. Johnny was an obedient youngster, willingly doing the farm chores asked of him. When Mr. McClintock died the farm was willed to Johnny, but it was to be held in trust by his widow until her death. It became known as the "Widow McClintock's Place" and proved to be immensely valuable when oil discoveries were made. The widow was shrewd enough to lease small portions of the farm, asking for only small amounts of cash for the leases but insisting upon a one-eighth royalty on all oil produced. When the avalanche of money began, the widow was able to buy a new cook stove, a silk dress and a small iron safe. After those few simple desires had been satisfied, Widow McClintock was content to sit on the front porch and rock. Occasionally she took time to tie up a package of currency with string and shove it into the small safe.

Johnny had no feeling of affluence either. He was busy working as a teamster, hauling barrels, tools or other supplies needed in the oil fields. On one trip, Johnny met and fell in love with Eleanor Moffit. They were married a short time later. As a wedding gift, the Widow McClintock gave Johnny a general store, fully stocked. This gave him an aura of respectability and stability.

Johnny was about nineteen, though he was far from mature. He had to buy supplies in Pittsburgh for his store, and he became entranced with the big city on his first trip there. He spent most of his time sightseeing, riding the horsecars and frequently enjoying a "festive drink known as a cocktail." Johnny became so engaged with the festive drinks that once, when he was taken in an alcoholic accident, he bought a barrel of syrup. Johnny was an addictive person, and he smeared syrup on his food, drank it like water and used it in various other ways. Once the barrel was empty, Johnny never touched another drop of it. But to his utter downfall, Johnny didn't acquire the same revulsion toward strong drink or his festive cocktail.

In March of 1864, the Widow McClintock died as a result of burns incurred when she started her favorite stove. The fortune then passed to Johnny. He wasn't quite twenty-one years old, and a guardian was appointed to conserve the assets until he was legally of age. But as soon as Johnny reached maturity the guardian was dismissed and Johnny looted the old iron safe. There is some question about the exact amount found, but according to *The Great Oildorado* there was $24,500 tied up in cash bundles. Although the amount was perhaps not as much as expected, it was certainly enough to give Johnny a substantial start. Oil was then selling for about $14 a barrel, and the widow owned several flowing wells. Johnny's immediate daily income became several thousand dollars.

Johnny was overwhelmed with his affluence, but a man named Dan Fowler appeared to help him through the financial quandary. Letting Fowler get near that amount of money was like inviting a weasel into a chicken coop. Fowler was a suave barrel salesman and was able to convince Johnny that he was one of the world's shrewdest financiers, but he needed a little guidance. Fowler would supply that. They took a trip to Meadville to survey some possible investments. There Johnny invested $69,000. He bought blooded horses, a farm in which to keep them, a business building and a residence, loaned a strange doctor several hundred dollars to complete a research project and gave a lawyer several hundred dollars for a few minutes' consultation on a trifling subject. Johnny needed none of these things, and it is reasonable to assume that Fowler was covertly paid a commission on each dollar frittered away on such foolishness.

Though, according to Johnny's own account, he had nothing to do except "loaf around, smoke cigars and watch my bank roll increase," he managed to meet another friend, William Wickham, who later became mayor of New York City. Shortly after their friendship began, Coal Oil Johnny visited Wickham in New York, where he whiled away his time playing pool, smoking his expensive Havana puro cigars, drinking wine, feasting and ostentatiously riding about in a polished, stylish carriage which was driven by a liveried coachman. In time Johnny brought his wife and infant son, Oscar, to New York, but they moved on to Philadelphia when Mrs. Steele became ill. Johnny left her there for a trip to Oil Creek.

Johnny had a faculty for meeting scamps, and at Oil Creek he was introduced to Seth R. Slocum, a man known to be as crooked as a ram's horn. Once Slocum realized that Johnny controlled a huge fortune, he never left the youngster's side.

Coal Oil Johnny refused an offer of one million dollars for the Widow's Farm, an astute move, but then he leased Wickham some of the best producing lands. Wickham saw that the lease terms were all drawn in his favor. The document was written in such a way that Johnny believed he could demand a million dollars whenever he wanted. But Johnny didn't read the fine print. The lease didn't provide for any such thing. Worse, on the strength of the arrangement, Johnny and Slocum returned to Philadelphia and rented a sumptuous suite at the Girard House, where the first of many baccahanalian revels began and lasted until the last of Coal Oil Johnny's fortune was exhausted.

Despite its immensity, Johnny's wealth was drained more quickly than anyone expected. Johnny gave Slocum a general power of attorney, which allowed him to draw upon Johnny's bank account, as well as charge anything, anywhere, for Slocum and his cronies. Not content with taking everything he could in a seemingly honest manner, Slocum steered Johnny into brace gambling games, where crooked dealers fleeced Johnny of whatever they could. It became a silent race to see who would get the last dollar, which was soon in sight.

Johnny was no whiner, and was a sport to the last dime. He bought a fine carriage, in which he stood up and threw money into the streets of Philadelphia, just to see the less fortunate scramble for it. When Johnny tired of that sport, he strolled the streets with five- and ten-dollar bills peeping from all of his pockets, his hatband and buttonhole. He seemed to sense majesty or grandeur as the gamins and newsboys snatched them away. Johnny pretended to fend them off with his gold-headed cane.

Some perverse trait made Coal Oil Johnny want to smash silk hats with his cane. He paid $10 for each crushed hat. A brisk market immediately developed in second-hand hats, many of which Johnny had already flayed, and hangers-on followed him to work his will for the $10 bill.

Following a petty squabble with a desk clerk, Johnny rented the Continental Hotel for a day at a cost of $8,000, just so he could fire the fractious clerk. Johnny was enraged when the hotel management rehired the clerk as soon as Johnny's one-day lease expired.

He was involved in a myriad of escapades and extravagances, so many that Coal Oil Johnny felt impelled to relate them later in an autobiography, which was privately printed at Franklin, Pennsylvania, in 1902.

Coal Oil Johnny had been living the wild life from his daily income, but as the wells petered out he began to sell his assets to continue his way of life. Eventually the assets all disappeared. By the summer of 1865, Johnny was broke and, of course, friendless. Slocum, like the other false friends, stole away when bankruptcy appeared on the financial horizon.

Coal Oil Johnny accepted the losses and his foolishness philosophically. He didn't complain, and apparently tried to be completely honest with himself in his book. "I spent my money foolishly, recklessly, wickedly, gave it away without excuse; threw dollars to urchins to see them scramble . . . was intoxicated most of the time and kept the crowd surrounding me usually in the same condition."

At the end of the prolonged debauch, Johnny wandered aimlessly about for a time, then returning to his wife and family, who had been living with her parents. His wife gave him enough money to buy a team and wagon, a status from which he had started. Johnny resumed work as a teamster, just as though nothing at all had happened. He quit drinking and smoking, joined a church and participated in civic affairs. While he remained in the Franklin area, the local newspaper commented favorably on his conduct, industry and change of ways.

In time, the Steeles moved from Granklin to Fort Crook, Nebraska, where Johnny worked as a railroad baggage agent often from 6 A.M. to 11 P.M. He died there on December 31, 1920, closing a fabulous era of the oil industry and one of its characters.

5

Pithole

As production increased along Oil Creek and other sections, there was a frantic scramble for barrels, containers of nearly any kind or transportation to get the crude oil to market. No one, certainly not those in the cooperage business, had foreseen such a wild demand for barrels. *Derrick's Handbook of Petroleum* observed: "The coopers could not make barrels as fast as the Empire Well alone could fill them."

Whisky barrels, nail kegs, even wash tubs were tried to contain the massive flow but it seemed nothing could hold back the tide of black gold. Oil was even run into earthen tanks, but this practice was quickly abandoned because of the fire danger and the loss from evaporation and seepage. Wooden storage tanks were constructed, but they too presented problems of leakage. It wasn't until some unknown genius suggested heating and treating both barrels and tanks with glue and similar substances that some of the problems were solved.

Oil was sold in barrels considered to contain forty gallons, but even then there were no exact standards. There was so much loss through leakage, and abundance of water and other impurities, that an extra two gallons was allowed in every forty in favor of the customer to offset the differences. This forty-two-gallon barrel was officially adopted by the Petroleum Producers Association in 1872 (and approved by Congress in 1916), but the standard had actually been set in 1866 by an unofficial agreement between the big producers in Pennsylvania which said in part:

"The present system of selling crude oil by the barrel, without regard to size, is injurious to the oil trade . . . as buyers with an ordinary size barrel cannot compete with those large ones. . . . From this date we will sell no crude by the barrel or package but by the gallon only. An allowance of two gallons will be made on the gauge of each and every forty gallons in favor of the buyer."

Even though the size of the barrels had been standardized, there was an even more desperate situation. How was this mountain of barrels to be moved to markets? Suitable transportation was almost nonexistent. The nearest railroad was forty miles away. Roads were little more than extended muddy or dusty ruts. Movement of the oil could best be accomplished along Oil Creek. There was only one obstacle. Except when Oil Creek was at flood stage, there was seldom enough water to float much more than a small boat of shallow draft. Shipments were made in rowboats and even canoes, but they were inadequate to meet the demands of the oil shippers. Flat boats, barges or barrels lashed together and guided by a man following in a rowboat were employed in Lower Oil or French creeks. Owing to the scarcity of barrels, attempts were made to fill open barges with oil; but unless these unwieldly craft were kept on even keel at all times, the oil tended to slosh back and forth until the barge capsized. The spilled oil would contaminate the water and become a fire hazard in the lower reaches of the creeks and finally in the Allegheny or Ohio rivers. Now that there was not so much difficulty in obtaining the oil, there seemed to be no adequate method of getting it to market and translating the product into cash. Finally someone came up with the idea which was blandly called a "pond freshet." It was probably the most unusual and spectacular method of transportation ever devised. It was an artificially created flood!

The scheme was simplicity itself. Oil Creek and its major tributary, Pine Creek, would be damned in several places, creating small mill ponds. When these miniature lakes were considered high enough and all of the barges were loaded with barrels of oil, a signal would be given and the dams broken. The resultant "freshet" would raise the water level by several feet as the water accumulated, enough to accommodate the heavily laden boats downstream, where there would be enough water to float the barges.

By mutual agreement, A. S. Dobbs was named Superintendent of Pond Freshets. He negotiated arrangements and fees with the landowners whose property adjoined the creeks, who were paid handsomely for use of their mill ponds, and levied taxes on each barrel of two to four cents for each one which reached Oil City. It was little enough reward to get the mill-pond owners to cooperate with "sailing" schedules. Miraculously, thousands of barrels managed to get through to Oil City, but the pond freshet system was little more than coordinated disaster.

The pond freshets were arranged about twice a week, depending upon the needs of the oil shippers, the amount of water available and the time required to clear away wreckage from the last sailing. There were any number of barges, ranging from two to eight hundred. Though figures differ, about 10,000 to 20,000 barrels usually made the journey, though nearly 40,000 barrels managed to get through during one spectacularly successful freshet in 1863.

Once the time had been arranged, the signal was given. Barge crews frantically pushed, dragged and pulled to get their craft out into the channel to obtain full advantage of the water rushing towards them. Many made it. Many didn't and were left stranded on a sandbar.

But once the oil fleet was launched, the dangers really began. The worst came when barges veered sideways in the current and following boats began piling up one upon the other, crushing, spilling and capsizing. The key to success was timing by the crew, and judging just when to push into the current, then how to keep the barge straight afterwards, was a delicate matter.

In some freshets, none of the boats got through to deliver their cargoes, and it was rare when a pond freshet was released without the loss of several boats and a great deal of oil. Herbert Asbury, in his fine book *The Golden Flood,* states that at least a million barrels of oil were lost in this disastrous method of transportation. So much was lost that many people made a living by simply skimming off oil downstream. But despite the loss of barges and contents, there was an amazingly small loss of life.

For those who managed to get through the wreckage and confusion and finally reach Oil City, which stood at the confluence of Oil Creek and the Allegheny River, there was another torrent awaiting the boatmen: cascades of whisky.

Safe arrival was cause for celebration. Survivors of the pond freshet needed something to quiet their nerves. Others celebrated because they had made enough skimming oil to afford relaxation. And the oil buyers dulled the senses of the boat owners with libations. The ensuing revelry gave the dishonest an opportunity too. Men would dive into the river, bore holes into the boats which carried oil in the holds and then go down river and skim off the oil as it floated past. Such practices brought about one of the great disasters in the history of Oil

Creek. A barge owner, believing that thieves were at work on his boat, peered into the hold while holding a lighted lantern. The resultant fire from this idiotic act destroyed forty fully laden boats, a bridge, several riverside wharves and many other structures.

Though fire, flood and ice gorges destroyed hundreds, maybe thousands of boats, thousands or hundreds of thousands of barrels of oil and other property, there was a poignant aspect to transportation which is generally overlooked.

The barges which survived the perilous pond freshet were returned by teams of horses, which splashed through the oily muck as they pulled the crafts upstream. This greasy mixture gradually ate away all of the hair from the horses' bodies until some were bald below their eyes. Suppurating sores erupted, and the lesions became larger. Painful, lingering death followed, and the life span of the stoutest horse was seldom more than six months after starting this gruesome task.

Transportation slowly improved when plank roads were laid, usable by teams, and railroads inched nearer the oilfields. Both means vigorously opposed a plan to lay pipelines which would help relieve the oil congestion. Pipelines would "only take the bread from the mouths of our children," the teamsters wailed at the threat. The powerful railroad lobbies applied pressure on legislators to prevent issuance of any franchises for pipelines, which owners would need to obtain right-of-way. But when both the emotional and political gambits failed and the pipelines became a reality, physical force was applied. Pipelines were pulled apart by teams, joints were quietly separated to drain oil without the owners' knowing and pumping stations were destroyed. There were numerous brawls, arson and sabotage, but, though many were injured, only one man was killed.

In 1860, General Samuel D. Karns, formerly of the U.S.

Army Corps of Engineers, was the proprietor of a salt well at Burning Springs which suddenly began to flow oil. The General proposed to build a six-inch cast-iron pipeline to Parkersburg. The ensuing bedlam of the Civil War kept the proposal from materializing, according to an account in *Oil Across the World,* by Charles M. Wilson.

Samuel Van Syckel put the pipeline business on a full-scale profitable basis by laying a two-inch crude-oil pipeline five miles long in the Pithole area. There was immediate and violent opposition from the teamsters, but the pipeline survived the contest and the character of the oil business was completely changed once again. Because the pipelines made carriage of more oil possible, and because of the expanded market, lease holders wanted to be sure they were getting all possible oil. Some speculated that mere pumping might not be completely draining the well. Oil was thought to be contained in "crevices" of rock, and "torpedoes" were brought into use. The device had its inception in water-well drilling and might have been applied successfully for oil.

Because H. H. Dennis believed the "crevice" or "pool" theory, he became convinced that an explosive lowered into an oil well and detonated would crack open any concealed pockets and release more oil, making it available to be pumped dry. His first attempt was a copper tube crammed with rifle powder and exploded underground with a fuse. Some water welled up, but no oil. It was a failure.

But the attempt pointed the way for others to experiment with explosives in their wells. The results were unsatisfactory, some of the explosions ending all oil issue. It wasn't until Colonel E. A. L. Roberts, a veteran of both the Mexican and the Civil wars, invented the "torpedo for artesian and oil wells," that success was obtained. His torpedo was a cast-iron flask filled with gunpowder, with percussion spaced at intervals within the container. The torpedo was detonated by a heavy cast-iron weight, called a "go devil," dropped from the surface and guided by a cable which secured the torpedo. Because of previous failures, oilmen were understandably reluctant to have their wells treated with explosive devices. But from Roberts' first test, on January 21, 1865, the torpedo was acclaimed a success. A new chapter in the infant, lusty industry had been written.

Roberts' torpedo worked consistently, and the idea was pirated by many other oilmen. But the Colonel was a hard-nosed man and inexorably pursued in the courts those who infringed on his patent. He is said to have filed hundreds of suits and forced many to desist as well as pay heavy damages for their infringements. Colonel Roberts is considered to have filed a record number of suits, the costs of which ranged, from various accounts, from $100,000 to $250,000.

Because of the dynamic growth of the oil industry, the choice Pennsylvania leases had been taken, and the rainbow chasers looked elsewhere for oil. One was the Pithole area, which was cut through by a small creek which ran roughly parallel to Oil Creek. Both were in Venango County in Pennsylvania.

The curious name Pithole originated from the odd rock fissures which were scattered along the rim of the narrow valley, through which the Pithole Creek flowed. Most of the "pitholes" occurred near the confluence of the Pithole and the Allegheny, near the tracts of land owned by the McCalmonts and the Widow Shaw.

Interest in these pitholes was provoked by the issuance of nauseous gases, which the pious said emanated from the fires of hell. Some even claimed they could hear the moans of sinners shrieking in torment from the pitholes.

Even some of the experienced oilmen must have believed there was some magic attendent to the area, its sights, sounds and smells. "Oil smellers" and "dowsers" formed a clan whose choice of oil sites was a result of supposed scents rather than science.

". . . The pretensions of the diviners are worthless. The art of finding fountains or miners by a peculiar twig is a cheat upon those who practice it, an offense to reason and common sense, an art abhorrent to the laws of nature, and deserves general reprobation," wrote Prof. Benjamin Silliman, Jr.

Perhaps so, but it was a dowser who discovered the first producing Pithole property on January 7, 1865, a 650-barrel-a-day well on the Holmden Farm, about seven miles from the mouth of the creek. The lease rights were held by I. N. Frazer and Tom Brown.

Finding oil was once again the least of the problems. Oil was selling for $8 a barrel. Teamsters were hired to carry away the accumulating oil, but their wagons became hopelessly mired. Many wagons weren't recoverable until summer. Finally some of the oil barrels were moved out on sleds, pulled by horses, and once these reached the market areas, word of the new strike exploded among the other oilmen. Leases immediately sold briskly for $2,000 per acre, plus royalties on each barrel of oil recovered.

Because the general public was slowly beginning to realize the incredible wealth which might be belched up from any one of the random holes drilled nearly anywhere, the Pithole discoveries caused a great deal more excitement than others. People who knew less about oil wells than a novice in a nunnery pressed their cash on promoters who promised much and produced little.

"Shares in wells were divided and subdivided until housemaids, cooks, porters and barbers boasted of 32nd, 64th and 128th interests in wells that never accomplished much more than to disturb the woodpeckers for a brief season," Asbury wrote in *The Golden Flood*. He added: "Tens and hundreds of thousands of dollars were paid for leases that could not have paid the purchasers ten cents on the dollar if every acre had yielded a thousand barrels of oil per day for a year. . . ."

Because of severe weather, the rush to Pithole and the new Oildorado area was delayed for several weeks. But once the tocsin is heard, the sound is never forgotten. Men and women sloshed through deep mud to get there. Pithole City was populated with 15,000 people almost at once. One hotel, the Astor House, was built in one day. A railroad was immediately graded to haul in both freight and people, and there were plenty of both. Banks, a newspaper, saloons, brothels, cooperages and warehouses appeared, all reached by streets which could be recognized by the three-foot-deep mud. Lots, sold for $10 before the boom, now brought thousands of dollars. A hotel built at a cost of $50,000 was sold three years later for $16 as firewood. Other buildings were moved in and out, but most of them were destroyed by fire, always the bane of boom cities.

Despite the fact there was no government, law, fire department, jail or any water system, Pithole somehow survived and flourished until 1868, when it was deserted en masse. Everyone there was bilious from the polluted water and the shroud of petroleum vapors which embraced the city, a condition which the Pithole press claimed made the city "distressingly healthy."

In time, families began to move in. There were a few "free and easy" women who were durable enough to follow men anywhere. In hearing one case, Judge Charles Highberger said:

"If I but choke my wife a bit or slap her only three or four times with nothing but an open hand, or belabor her over the back with a barrel stave for a Sunday diversion, she seems to get kind of stiff and tottery, and goes huffing and clucking around for a whole week at a time. . . ."

While the women may have been frail and weak in one man's opinion, others were equally odd. Someone described the lease dealers, often called "underground farmers," as looking like "the lower class of scarecrows under the influence of a galvanic battery." Whatever they looked like, many of their customers looked woebegone when they found their Pithole oil shares to be worthless. The stranger who appeared with a smile on his face, a note of confidence in his voice and larceny in his heart often sold, say, sixteen one-eighth shares in a non-existent well and then left Pithole hurriedly. Wells were promoted by scattering oily sands about derrick platforms, dumping barrels of crude oil down the hole or even pumping oil from concealed underground tanks. Leases were sold and resold until the share prices reached astronomical, completely unreasonable figures. Even then many of the lease rights were never exercised, nor the wells drilled. It is possible that more money was made by these oil swindlers than was ever pumped up from Pithole wells.

". . . This is the way that money is made at Pithole," the Oil City Register commented on the sharp, devious practices. Unfortunately, many of these nefarious schemes, born of the earliest oil days, prevail.

During February, 1866, a fire ravished Pithole City, consuming many of the major buildings. Water used to fight the fire was drawn from the primary and secondary water tables, and oil was flushed from wells which had previously issued only water. Widow Rickerts, for one, scooped up more than one hundred barrels, even while she was refusing proposals of marriage from self-seeking suitors. Others tried to pump their wells to obtain oil. Very little oil was obtained, but the water was so drained from the water levels that it had to be brought in by stoneboat for some time afterward. From that time on, water was always scarce and fires were usually left to burn themselves out. But even before the ashes were cold, new leases were being sold and new wells being put down.

The Pithole oil bubble burst almost as suddenly as it had been inflated. In 1867, oil dropped to $1.75 a barrel for forty-two gallons, a poverty level. Wells were shut down, drilling ceased and companies went broke. Production stopped. Walls were papered with now worthless stock certificates or used to kindle fires. One man pasted $53,000 on his walls.

People began to drift out of Pithole, until the movement became an exodus in 1868 and the city was completely deserted. Most of the men moved temporarily to Shamburg, to Pleasantville and later to the rich Bradford fields.

Today nothing remains of Pithole except memories.

6

A Giant Is Born

Most men investigated the oilfields with an eye to obtaining a successful lease or getting in on some aspect of the producing end of the oil business. Not so with John D. Rockefeller, a twenty-year-old Cleveland, Ohio, produce merchant who was fascinated with the fact that a gallon of crude oil refined into kerosene sold for twice as much as the entire barrel of oil from which it came. With that in mind Rockefeller visited Titusville in 1860 to survey the possibilities of an investment in the oil industry. For two months he looked, asked questions and studied the situation. It was immediately apparent to this shrewd young man that a fortune could be made if control of the refining and transporting of oil could be achieved, but not in the extremely risky business of drilling a well. There was scant risk in refining the products which other men risked their fortunes and lives to obtain. For nearly two years, Rockefeller considered the matter, leaving no detail unattended. This attention to detail was one of the keys to Rockefeller's success. Though he was already a modestly wealthy young man, dealing in meat, grain and produce, Rockefeller always sought to improve his situation. His elder sister, Lucy, once observed:

"When it rains porridge, you can be sure that John's dish will be right side up."

Rockefeller had been intrigued with the excitement generated by the Drake discovery well, though he wasn't sure that

Drake had "tapped the mine." And Rockefeller seldom took a step without having tested the ground first, preferably at someone else's risk.

Sam Andrews was a man of Rockefeller's own cut. Andrews was working for a refinery, hoping to get enough money with which he could build his own. Andrews was finally able to convince Rockefeller and his partner, Maurice B. Clark, to invest even though the oil market was depressed at the time. Oil was selling for only ten cents a barrel, with no relief in sight. But Rockefeller, a man who "paid attention to details," believed that if an investment was sound, it should be embraced in good times or bad. If there was any doubt, avoid it like the plague at any time. As a result, the company of Clark & Andrews was formed, Rockefeller refusing the use of his name until complete success had crowned their efforts.

The refinery at first produced ten barrels of refined crude oil each day, but that figure was gradually increased to one hundred barrels. They prospered as railroads made movement of crude oil easier and cheaper and the demand rose for the improved refined products. Rockefeller believed that Cleveland would become the key point for the accumulation and refining of oil, with trans-shipment to eastern markets. This judgment, among others, caused dissension between the partners: Rockefeller and Andrews against Clark. No compromise could be reached, and so the firm was dissolved and put up for auction to the highest bidder, winner take all.

It was a tense meeting which convened in the offices of Clark's lawyers. The men bid quietly, each trying to sound out the extent of his final bid. Their resources were about equal, but Rockefeller's determination and financial courage were greater. He finally bought out Clark for $75,000 and a half interest in the grain business. This was the first giant step towards a fortune which would later be counted in the hundreds of millions of dollars. Rockefeller was in full command of what would become the world's greatest trust, the Rockefeller & Andrews Company, soon to emerge as The Standard Oil Company.

Rockefeller's judgment of men was usually astute. Andrews proved to be an excellent production man, often coming up with new ideas to produce kerosine more efficiently, as well as other efficiencies. Such improvements were a delight to Rockefeller's heart. When they considered the price of barrels to be too high, they bought a cooperage plant and produced their own barrels. Instead of buying oil from jobbers, Rockefeller dealt directly with the producers, eliminating the commissions. Transportation was expensive, so Rockefeller purchased teams and wagons to haul his own crude oil and finished products. As Andrews improved refining processes, their kerosine was on sale everywhere, undercutting competitors in price and quality. With success seemingly assured, Rockefeller convinced his brother William to sell his produce firm and join with them. He was to become the representative for Rockefeller & Andrews assigned to handle sales and distribution in both the eastern (ostensibly New York) and European markets. He agreed, and the William Rockefeller Company was formed. It became the first major competitor to the refineries, which heretofore had a monopoly on the trade in these eastern and European markets.

It wasn't that Rockefeller objected to monopolies, because he firmly believed they "brought order out of chaos." But he did indignantly oppose any monopoly which he didn't or couldn't control. He intended the only refiner and dealer in oil products to be Rockefeller and his associates.

The mark of a successful executive is a man who can select

able assistants, trust them and delegate authority without any reservation. Rockefeller was a master in such judgments, as was brilliantly displayed when he chose to make Henry M. Flagler a fourth partner in the growing Rockefeller enterprises.

Flagler, a runaway at fourteen in 1844, took nothing with him except a nickle, a coin which he still had when he died, worked at numerous jobs, succeeding at each one and bettering himself with each change. Through remote business dealings, Flagler and Rockefeller became acquainted, and each admired the other's abilities. Rockefeller, usually withdrawn with strangers, became openly friendly with Flagler, and he was Rockefeller's next choice to join the firm as it expanded. But since Flagler had little money, Rockefeller suggested that he convince a whisky-dealing relative to invest and then have Flagler represent the interests in the Rockefeller & Andrews Company. It worked, and Flagler became associated. It was a sympathetic choice. Flagler had a sound, logical mind and an uncompromising attitude toward competitors, making him the type of man Rockefeller could entirely understand and appreciate. The firm was reorganized under the name of Rockefeller, Andrews & Flagler, with the William Rockefeller Company remaining in New York to handle the export and other trade. Another refinery was built, and the incubus was about to settle about the oil industry.

For once it appeared that Rockefeller was on the short end of the odds. Rockefeller and other refiners in Cleveland were at a disadvantage in competing with the eastern refineries and marketers. The Cleveland dealers had to make two shipments, first shipping the crude oil from the wells and then shipping again after refining. Other refiners made only one, which slashed their costs to the point where Cleveland refiners couldn't successfully compete. But when the second Rockefeller refinery was built, he became the largest producer and believed they could now deal advantageously with the railroads. Flagler proved his worth. With the implied threat that all of the Cleveland refiners would combine and ship products by the slower but less expensive boats, Flagler obtained a rebate on every barrel of crude oil shipped to them, and an even larger rebate on the kerosine which was shipped from Cleveland to New York. Months later, an even larger concession was obtained; but for this Rockefeller, Andrews & Flagler promised to do their own loading and unloading operations, no longer require fire insurance and promise to ship a minimum of sixty carloads a day.

The hard-knuckled dealing with the railroads enabled the Rockefeller interests to ship kerosine to New York and export markets cheaper than anyone else because of his more efficient methods. At that time there was nothing illegal about secret rebates. They were allowed in other industries. Rockefeller and Flagler simply showed the way for the oil industry. Rockefeller's was the first combine large and bold enough to demand enormous concessions from railroads.

Rockefeller moved ruthlessly into the eastern markets, using the advantages of size and economics to force one refiner after another out of business. By 1869, the firm of Rockefeller, Andrews & Flagler was probably the largest refiner in the world. As profits blossomed, they bought tank cars, refineries, storage tanks and teams, and established a research laboratory, the first, to create new products from their petroleum, either crude or refined.

It was on January 10, 1870, that the Standard Oil Company of Ohio was organized. It became perhaps one of the most maligned corporations in all of American corporate history. While Rockefeller always maintained that he was working for the over-all good of the oil industry and mankind, there were many so unkind as to doubt his basic motives. Oilmen who had been defeated and crushed by the Rockefeller interests were understandably bitter, but they also realized that the Standard Oil wasn't all bad:

". . . If a man in Oildom drilled a dry hole, backed the wrong horse, lost at poker, dropped money speculating, stubbed his toe, ran an unprofitable refinery, missed a train or couldn't maintain a champagne style on a larger-beer income, it was the fashion for him to pose as the victim of a gang of conspirators and curse the Standard Oil vigorously and vociferously," John J. McLaurin wrote in his *Sketches in Crude Oil.*

During the early 1870's, the oil market was glutted with an estimated five million barrels in storage. There were not enough machines or lamps to consume that amount of oil, either crude or refined. But the oilmen, fearful that someone would steal the market, continued to drain their wells of all the oil they could. Some producers tried to reach agreements to control the output, but greed limited observance to a few days. The market palled, and refineries began to close. At the time, Cleveland was still the center of American refining, the plants there being able to refine more than the full production of all the oilfields and produce enough refined products to supply the entire world. Of these, Standard Oil Company of Ohio was the biggest and the most efficient. Rockefeller didn't quail or lose his courage in the face of the bleak outlook and glutted market. Systematically, Standard Oil began to absorb the failing plants; few would have survived anyway.

Not content with the incursions that Standard Oil had made

in controlling the refining industry, someone suggested formation of the South Improvement Company, which would tie up the railroads, barrels, storage areas or any other loose ends which might threaten the proposed monopoly. Weaker opposing companies would be squeezed out. Unfortunately for success of the scheme, the plans were publicized before their execution, and determined opposition sprang up. The independent companies realized that it was "one of the most gigantic and dangerous conspiracies ever attempted."

Perhaps it was. Secret agreements controlled the shipments of oil, giving equal shares of the business to cooperating railroads. Rates for competitors were to be doubled. Rebates were to be increased because of the increased rates for those outside the combination, and all information about the business methods of competitors was to be made available by the railroads.

The South Improvement Company's plans, however, were checked by the people of the oil region and by Congressional investigation.

Rockefeller hedged his part in the failure by saying later that he never believed the scheme would work, even though the target and method strongly hint of his rapaciousness. Whether or not Rockefeller can be held directly accountable for the company, he and Flagler did own shares which were never paid for because the company overreached itself and folded before the death grip had seized on the oil industry. Fear, hatred and resentment consolidated against the South Improvement Company. Rockefeller never fully understood the independent nature of the American people. With the formation of the Standard Oil Trust, Rockefeller said: "The day of the combination is here to stay. Individualism has gone, never to return."

Rockefeller reckoned without the men who refused to pay tribute to anyone. Mass meetings were held, the independents agreed that no oil would be sold to the South Improvement Company and plans were drawn to build independently owned refineries. Political pressure got the charter for the company revoked, and the threat was removed. At least a portion of the threat: the Standard Oil of Ohio still remained in control.

With Flagler and William Rockefeller, capitalization of Standard Oil was increased, and the trio moved to buy or acquire all of the refineries in Cleveland. While the plan might seem too large for even Standard Oil, Rockefeller bargained. "Sell or be ruined" was the usual offer.

With control of railroad rates, rebates and cheaply produced kerosine, the refineries threw in the sponge. Rockefeller then had the properties appraised by his own men, offering cash or shares of stock in Standard Oil. He urged acceptance of the stock, and those who followed his advice became wealthy men. By March, 1872, Standard Oil controlled Cleveland's refining. Oilmen suddenly realized that Standard Oil had simply replaced the South Improvement Company. Rockefeller moved to obtain oil-well production through various secret agreements, to keep prices firm and then form an association with refiners in other areas to cooperate in keeping prices at a high level. Because of general distrust of Rockefeller interests, the plan failed. Though it wouldn't have mattered if he had, Rockefeller never understood the hatred and resentment which was felt about him.

Rockefeller was neither dismayed nor discouraged. He was determined to control the entire refining industry, and he secretly set about the task. In a conspiracy with the Lake Shore Railroad, Standard Oil got huge rebates and vital business information about his competitors.

He maneuvered from a position of strength, increasing the capitalization of Standard Oil to $3.5 million; then he made deals which proved to be of immense importance, just as Rockefeller knew they would.

One such deal was with the Erie Railroad, which leased him their shipping facilities at Weehawken, New Jersey, for a very modest sum. The costs for shipping were set at the minimum for Standard Oil. Although this was done before Rockefeller became involved in the United Pipe Line systems, another important advance for Standard Oil, the Erie Railroad agreement was not a power play. The depression of 1873 made all railroads desperate for cargoes, and the Erie was trying to outbid the Lake Shore arrangements. The New York Central was convinced to make equally advantageous deals with Standard Oil, and when the United Pipe Lines was taken within the complex transportation arrangements, the future looked very prosperous. Any company, including the railroads, which even hesitated in accepting Rockefeller's loaded blandishments were threatened with starvation through control of oil and other shipments. Only two companies dared to fight Rockefeller rather than meekly accept defeat.

One was the Pennsylvania Railroad, (the "Pennsy"), and the other the Empire Transportation Company, usually called the "Green Line." Colonel Thomas A. Scott headed the Pennsy, while Colonel Joseph D. Potts presided over the Green Line. Both men watched Rockefeller's moves apprehensively, knowing it would be only a short time before they would have to fight for their financial lives. But they weren't afraid, and the battle lines were drawn for a heroic struggle between financial giants.

Potts and Scott began buying refineries, warehouses, cooperages and marketing facilities. They moved to organize the oil producers and helped escape throttling by the Standard Oil. Potts lobbied to break Rockefeller's monopoly but found that the legislators were less than sympathetic, no doubt because of pressure from Rockefeller. Potts observed that the legislators had a record of failure and disgrace.

"If it has taught us anything, it is that our present lawmakers are, as a body, ignorant, corrupt and unprincipled," he observed bitterly.

When the Pennsy and the Green Line struck back, all shipments on these lines were banned by Rockefeller. The New York Central and Erie railroads joined with Standard Oil to help put down their resistance. They hampered shipment of freight, lowered rates and hiked rebates. Rockefeller moved into Scott's and Pott's markets, underselling them at every point while still making a profit for his companies. Scott and Potts lost millions, while Rockefeller became richer.

The rate war raged for five months. Then, when the bloody railroad strikes of July, 1877, erupted, the end came quickly. Rioting, burning and killing—forty-five men were slain—and huge property losses forced the complete surrender of the Pennsy and Green Lines. Curiously, Rockefeller was magnanimous in their defeat. He perhaps secretly admired those who had the courage to oppose him. With the consent of the Pennsy, Green Line, Erie, New York Central and the Baltimore & Ohio Railroads, the Standard Oil was appointed as "evener" to distribute the oil shipments among all of the various lines. Huge rebates were allowed the Standard Oil and its various subsidiaries. Standard Oil also bought up all of the refineries, tank cars, pipelines and supporting equipment with which Potts and Scott had tried to defeat Rockefeller.

In 1878 John D. Rockefeller, at thirty-eight years of age, stood in almost complete control—about 98 per cent—of all American refineries and pipelines.

The giant—Standard Oil—had been born and baptized with oil, and the industry would never be the same.

7

California's Black Gold

Except for the laziness of some muleteers, who dumped one hundred casks of oil off in the Panamanian jungles, history would probably record that the first crude oil came from southern California instead of Titusville.

Four years before Colonel Drake began his historic drilling, George S. Gilbert, a former dealer in whale oil, gathered crude oil on the apron of Sulphur Mountain, not far from Ventura, California. In a primitive refinery, Gilbert boiled off some of the impurities from the viscous, black crude oil. He sold some of it for axle grease and the rest to be used in lamps, when it would work.

Gilbert was thinking in larger terms than California, however. In time, he accumulated a hundred casks of refined crude, and at his request they were dispatched to A. C. Ferris of Brooklyn, who believed he might establish a brisk market in New York. Sent by water, the oil had to be unloaded and shipped across the Isthmus of Panama. It was a cumbersome cargo, and the disgusted muleteers discarded the casks in the jungle, where they were found many years later.

Gilbert saw an additional opportunity when he heard that General Andres Pico and his son, Romulo, were dipping up oil from seepages in the canyons north of the San Fernando Mission. The Picos used a crude still to process the oil; then they sold it to the mission padres. The Franciscans used it for healing, lubricating oil and some illumination. By modern standards, the oil would be considered of poor quality, viscous, black and smelly.

By this time, California was in full grip of the gold fever and no one was much interested in searching for or developing any oil deposits. Gold was the thing. There was no need for expensive tools, rigs, refineries or marketing facilities. Dig gold out of the ground, and it was instant cash. Oil was something else. No one could envision that more than three thousand by-products would ultimately be taken from crude oil.

One of the California oil pioneers came from the Titusville area. He was Lyman Stewart, who was born in Cherrytree Township in 1840. Stewart was literally born to the oil business; his birthplace was only ten miles from Titusville and the Oil Creek region. He was used to the "dowsing" method of seeking oil, and a "nose for oil" was the only method known of seeking a site to drill a well. Nothing was known about oil geology.

Stewart was forty years old by the time he arrived in California, a veteran of three years' service in the Civil War, a father, highly religious and the inventor of several types of oil-well drilling tools. He was nineteen years old when the oil fever enthralled the Venango Valley. Stewart was living with his thrifty Scottish parents and was apprenticed to his father as a tanner. Young Stewart thoroughly detested the trade, but it proved valuable to him in a quite unexpected way. In the course of riding throughout the Valley collecting raw hides or delivering them after tanning, Stewart learned where every oil spring or seepage was located, though he paid scant attention to them at that time. Everyone knew there was oil in the area, but other than minimal use as Seneca Oil, no one sought the "ugly grease."

No one, that is, except Timothy Allen, who in 1820 remarked that the grease might be collected and "become a profitable article of commerce." But Allen was busy with his duties as president of Allegheny College and didn't follow his own advice. Allen said:

"Fifteen barrels were once taken in one season from a single [oil] pit. It was sold for $2 per gallon but the common price is now $1.50. It is one of the most penetrating liquids in nature. No wooden vessel is impervious to this; even a glass in which it has stood for sometime cannot be cleared of the scent. . . . This oil is much esteemed for its efficacy in removing rheumatic complaints. It burns well in lamps, and might be advantageously used in lighting streets. If by some process it could be rendered inodorous, it would become an important article for domestic illumination. . . ."

Though he had read the comment, Stewart remained unimpressed and disinterested even when Venango County swarmed with oil seekers. By that time Stewart had decided to become a Presbyterian missionary, not only because of his profound religious beliefs but also because that profession would allow him to escape the wretched tanning business. At some point, Stewart invested his bank account of $125 in a one-eighth share of a lease located on the John Benninghoff farm. Unfortunately, none of the eight enthusiastic partners had enough sense to retain enough money for drilling and development after they had paid $1,000 for the lease rights. Their money was forfeited. Six years later, the area Stewart had selected to drill brought in a three-hundred-barrel-a-day well.

Benninghoff, a tight-fisted farmer of German extraction, shrewdly leased his land in small parcels which brought him as much as $6,000 a day during the height of the oil excitement. He wasn't as shrewd about banks. His royalty and lease money was kept in a flimsy old safe. Four robbers stormed into his house one day and held the entire family at pistol point while they extracted $200,000 from the safe, making it one of the biggest and most successful robberies of that time. None of the money was ever recovered.

Though Stewart had lost all of his savings and had to take up tanning again, he wasn't completely discouraged. He saved his money, determined to make another try at the oil business. It wasn't that Stewart was so entranced with the oil business per se as much as he was determined to succeed where he had once failed. With enough money saved, Stewart with some others leased a portion of the Boyd farm and brought in a small, steadily producing well. But the price of crude oil had dropped to pennies a barrel and Stewart couldn't even afford to pump the oil they had located. Once again, Stewart lost both his lease and his money.

In previous years, the depressed prices for crude oil had ruined many investors, especially those with limited capital who couldn't weather a prolonged financial drought. But the pioneers had shown the way, and Stewart and the others understood the values now. He was willing and determined to take another chance despite failures.

Upon Stewart's return home after his service in the Civil War as a cavalryman, he was startled to find that the village of four hundred people he had left now was a city of six thousand. Though he felt an obligation to his parents, Stewart was still determined to get away from the hateful tanning business. After a trifling amount of business school education, Stewart began dealing in oil leases. He had some shrewd deals, which gave him enough capital to think again about making personal investments. He bought wisely and cautiously. Stewart bought small shares, sometimes as little as one-sixty-fourth. He spread his investments over many leases, rather than backing only one lease. His capital was spread thinner, but the chances for smaller, surer success were greater. Stewart pursued the practice and prospered.

By 1872, Stewart had amassed more than $300,000 in the bank, and had an income estimated at $50,000 a year. Success finally seemed assured, but Stewart, ebullient and flushed by his achievement in the oil industry, assumed that he was equally infallible in others. Dabbling in agricultural machinery, Stewart went backrupt. He had to start all over again, once more. Discouraging.

On the financial bounce, looking for another opportunity, Stewart fortunately met Wallace Hardison, a banker who was willing to advance the cash in exchange for Stewart's know-how in the oil business. The agreement was sealed by nothing more than a handshake and never otherwise formalized by a written contract until their combined assets were so large as to be unwieldy with only an oral agreement. And, despite their close association for many, many years, Stewart and Hardison always addressed each other as "Mister" instead of using their first names.

Their partnership got off to a good start when Stewart invented some oil well tools which sold briskly. They were also in on the beginnings of the vast, rich Bradford Field, which in 1881 produced an estimated 80 per cent of all American oil. Unfortunately, though, it was also a time of the emergence of the giant Standard Oil, which held a death grip on the arteries of the industry by capturing the refining and transportation systems which served the oil industry. Prices were forced below pumping costs—as little as eight cents a barrel for crude oil.

This time, though, Stewart didn't go broke, but he did decide that he'd had enough in the Oil Region. He chose to go west. Hardison and Stewart sold their assets for $135,000. Leaving his wife and other children behind, Stewart with his fifteen-year-old son, William, headed toward California in 1882. As always, Stewart carried his Bible with him.

In the previous twenty-five years, California's oil production had been only a half-million barrels, less than what the Bradford Field gushed forth in one week. However, the prospects for the future seemed good. Two railroads were engaged in a rate war bringing passengers from the east to California for as little as $1. Experts saw that California's population would double in two decades. That meant an active oil market.

Stewart approached the new venture with caution. He was now a wary investor. He studied what had been accomplished in California and what was being planned to exploit oil. Stewart discovered that even before the gold rush, which began in 1849, Thomas Larkin, the American consul to the Mexican government at Monterey (not far south of San Francisco), reported that bituminous pitch had been found throughout California and was being used to waterproof roofs. Larkin estimated that it was to be had in commercial quantities, and with the potentially expanding markets, the product was desirable.

Stewart also learned of Gilbert's earlier attempts to gather and refine the crude oil in southern California, distilling his "liquid bitumen and asphaltum" on the Ojai Rancho.

Stewart read the enthusiastic reports of Prof. Silliman, Jr., the scientist who had also examined eastern fields. If anything, Silliman was more excited about the California prospects than he had been about eastern deposits. Most of his interest was centered on the seepages along the San Antonio Creek and Sulphur Mountain.

"The property covers an area of 18,000 acres, on which there are twenty natural oil wells, some of them very large," Silliman wrote to Thomas Scott on July 2, 1864. "The oil is struggling to the surface at every available point, and is running away down the rivers for miles. Artesian wells would be fruitful along the double line of thirteen miles, say for about twenty-five miles in linear extent. The ranch is an old Spanish grant of four leagues of land, lately confirmed and of perfect title. It has, as I have stated, 18,000 acres. . . ."

Professor Silliman estimated ten wells would bring a net profit of $1,365,000, a figure which seems to have been based more on his imagination than on reality. It was big talk for a conservative Yale professor, especially when it was later disclosed that Silliman's investigations had consisted of little more than a rambling stagecoach trip through the various areas.

However, Silliman's friend Scott, who, with Carnegie and some others, had made a fortune in Titusville, was anxious to get in on the start of what was expected to be another California boom. First gold, now oil! Silliman assured Scott that there was more oil in the California hills "than all of the whales in the Pacific Ocean." Though ordinarily astute, as indicated by his having been president of the Pennsylvania Railroad and Assistant Secretary of War in Lincoln's cabinet, Scott overlooked the limited western market and the feeble transportation system connecting it to the densely populated eastern areas. He immediately formed a syndicate to lease more than 300,000 acres of oil land, ranging from Los Angeles in the south to Humboldt County on the far northwest coast of California. Three separate companies were formed, with a total capitalization of $25 millions, most of which was "blue sky" promotion—that is, stock issued on expectation and hopes rather than real assets.

Tom Bard, who was later to become president of Union Oil, D. C. Scott and J. A. Beardsley were sent to California to begin drilling. The first steam-powered drilling rigs were purchased and shipped around Cape Horn. Before they arrived, Colonel E. D. Baker was punching holes in the Los Angeles area, a site which later became known as the Miracle Mile. But Baker had little success. The first oil well actually drilled in California was on the Davis Ranch in Humboldt County during 1861. It, too, was a modest success. Most of the oil was recovered from adits, one-ended tunnels which were pierced into eminences, allowing the oil to drain off by gravity.

Bard unloaded his tools, boilers, engines, furnaces and retorts at San Pedro, and then hauled them over mountain and valley about one hundred miles to Ojai Valley, ironically passing over many vast underground reservoirs as he plodded along.

As Bard prepared to spud in his wells, three San Francisco merchants formed the Union Mattole Oil Company, directing their interest in the Mattole River area of Humboldt County, about fifty miles south of Eureka. Oil seepages were abundant there, and they were nearer to San Francisco than the southern California deposits. San Francisco was the principal market, and transportation there would be less of a problem if the wells proved up.

"The first shipment of coal oil from Humboldt County was made on Wednesday last," the *Humboldt Times* reported on June 10, 1865. "Mr. F. Francis of Ferndale brought into town six packages of from fifteen to twenty gallons each of coal oil taken from the well of the Union Mattole Oil Company. This will go to San Francisco by the present trip of the steamer, and is the first oil shipment of crude oil from the oil regions of this area. . . ."

This marked the first shipment of crude oil from a drilled

well in California, previous oil having been gathered from oil springs or seepages. The Union Mattole oil was sent to the Stanford brothers' camphene refinery in San Francisco. The Stanfords, all of whom were to become influences in California affairs, had been secretly buying up Union Mattole Oil stock, but their optimism was unjustified. The first well in the Mattole area proved to be the best one, and even that wasn't very much. The others were dusters, or unprofitable—if they produced any oil at all. However, the modest success of the Mattole River wells was enough to touch off an oil boom which continued to blossom for a couple of years. It brought about the expenditure of $1 million to produce a mere 5,000 barrels, which were sold for a total of $10,000! Hardly a black gold bonanza.

One Mattole River producer transported oil in small containers by mule over the thirty-odd miles to Centerville, from where it was shipped on wagons toward Eureka, another twenty miles. The oil was then loaded on a steamer bound for San Francisco, 250 miles to the south. With minute production and wobbly and expensive transportation, California producers couldn't compete with the eastern refined oil, which sold for 54¢ a gallon. To complete the debacle, which highlighted the earliest days of the California oil industry, the federal government on March 17, 1865, withdrew all oil lands in Humboldt County from further exploration or exploitation, declaring them to be agricultural lands, and thus placing a cloud on all land titles in the affected area.

Tom Bard, Scott and Beardsley weren't concerned with the woes which plagued others. They had 227,000 acres in which to drill. The first hole was put down in a brea ("tar") pit, along the San Antonio Creek (later renamed the Ventura River). When they found only a trace of oil at five hundred feet, the well was abandoned. Another hole was put down. It was dry. A third went to 520 feet before they decided it was a duster, and quit.

By this time, the drillers, especially Scott, were becoming somewhat panicky. He hired two doctors, John Torrey and Charles Jackson, both skilled mineralogists, to examine their operations and try to solve the problem. After all, Silliman described this area as having oil "flowing down in rivers." Together, the men chose another location. That well, too, was dry.

The fifth well, punched down by jigging, brought six barrels a day at one hundred feet, an amount which was commerically worth while.

The one successful well had been extremely expensive, but Bard had learned a great deal. The clayey sedimentary soils and strata of the eastern Oil Region were one thing to explore and drill. But the flinty, rugged and faulted formations in California presented problems not previously faced. The strata were convulsed, upthrust and, in most cases, exceptionally difficult to pierce. Bard believed that oil would be found in "pockets" within these folds. This was the beginning of the "anticline" theory, which was later established as being essentially scientifically correct. He also believed that oil would be found at deeper levels than those in which they had already drilled.

Without further fanfare, Bard began to drill another well to test his theories. He hurried with the work because he feared that apprehensive, timid stockholders would demand that work be halted before another expensive duster was drilled. Bard proved that he was partially correct. At least he brought in a twenty-barrel-a-day well at 550 feet, which at that time was the best in the west.

But there was trouble ahead.

Just when the oil was discovered and the prospects bettered for more, the market for oil virtually disappeared, at least for the western operators. Oil shipped in from the east undersold California crude oil. Besides, a depression followed the Civil War and money, by late 1866, was scarce. Bard had spent about $200,000 in drilling six wells, only two of which were worth while (though that is a remarkable percentage in light of modern exploration success). But even if all of Bard's wells had been big producers, what now could be done with the oil?

As expected, the orders arrived stopping all further work. Others closed down too, all except the Stanfords with their Union Mattole Oil Company, who said they were compelled to do some token drilling "to safeguard their titles."

It seemed that by the close of 1866, the California oil boom had come a cropper. The bubble burst. Only time would tell!

8

Doodlebugs and Dowsers

Because the occurrence and formation of oil was so little understood. some of the early theories as to how it should be sought and extracted seemed touched with sorcery or drawn from alchemy.

And, because the first oil had been found along creeks or springs, the "creekology" theory appeared. It was believed that an underground stream of oil flowed along, following the watercourse, both seeking freedom in the oceans. Both the oil and the water were constantly renewed by drainage from the slopes, the oil coming from where coal deposits lined the banks. Oil did occur occasionally in conjunction with coal, and "coal oil" was extracted from a certain type of rock. Other than this, the creekology idea had scant basis in fact, but that didn't impress anyone who had once found oil near water. Oil was where they found it.

As the search for oil became more determined, other ideas were advanced. Some men examined the surface indications, such as the various plants being present or absent. Prospectors noted that oil came up with brine wells, and it was believed that these towers or domes of salt would prevent or encourage certain plants. This observation about salt domes was remarkably accurate, though not for the right reasons.

A touch of superstition was apparent in the idea that graveyards were usually productive, probably because of the magical aspects imputed to oil by the ancients. The cemetery search had some merit, but not because of the spirits. Graveyards are usually located on high ground, and such eminences often indicated anticlines or curved formations below, which occasionally contained oil.

One wildcat driller, apocryphally called "Hardboiled," spudded in a well near a cemetery, poured the cement to set in drilling tools and pipe and then waited for it to harden. While he was gone, some puckish drillers nearby drove a pipe on an oblique angle so it would intersect Hardboiled's cellar. When he returned, they moaned and set up ghostly voices. In sepulchral tones they pleaded for Hardboiled to allow them peace and rest and no more drilling in their graves.

"Have mercy on me," the driller yelled. "I quit." He fled the area; later the site brought in a fine well.

Oil occurs in four types of rock. Most common is the formation called the "anticline." This is layers of rock bent in undulating curves, resembling several inverted bowls. At the top of the anticline is gas, then oil and then salt water, if there is any. Directly above this anticline is a cap rock, an impervious layer often composed of limestone which keeps the oil from further migration. As noted, there are no caverns, crevices or pools of oil as such—the porous rock holds the oil, like water in a sponge. The oil is withdrawn when the drill reaches the critical area. Of course, not all anticlines contain oil, which is one of the reasons that millions of dollars are spent each year in the frantic search for oil. Even when an anticline is located after considerable searching, there is no guarantee that oil will be found within it.

Another type of oil formation is a "fault." This occurs when a horizontal layer of rock breaks under pressure and one section moves up, the other down. Many times oil-bearing rock is trapped in the movement under the cap rock.

A third type of oil occurrence is the "stratigraphic trap," which is essentially another underground movement of strata, trapping oil from further migration.

The fourth is the salt dome, an intrusion of almost solid salt which is forced up through layers of rock, forming traps on its sides as it pushes upward under immense pressures. Not only are oil traps formed along the sides, but often at the top, making these salt domes very significant in the successful search for oil. Sometimes, these domes can be discovered from mounds or humps on the surface.

Petroleum geologists have finally come into their own, using the most modern methods to discover underground reservoir rock. Seismographs, the instruments used to measure the occurrence and intensity of earthquakes, are employed extensively to map strata of rock. When oil geologists use these delicate instruments, small explosive charges are set several feet underground. When they are fired, miniature earthquakes are created. The vibrations are recorded on the machines' revolving drums. Seismic earphones give the geologists at least a strong indication as to what type and formation of rock lies below. Dense, tough rock reflects vibrations, while others tend to absorb the shocks. This is now the most important method of locating oil deposits on land. The most important offshore method is vibresis, which, to protect marine life, uses electronic vibrations to indicate formations.

The force of gravity is measured to assist the geologist in choosing likely sites for drilling. This is called the gravemetric method, and experts, by noting changes in the pull of gravity, can deduce what might be underneath them.

Magnetic measurements are also used, information which all adds to the geologist's total store of information by which he can make a sound decision. Occasionally these magnetometers are trailed from airplanes, a method which covers an immense area quickly and economically. These are used in conjunction with aerial mapping, pictures being taken with highly sophisticated cameras which show outcroppings of rock in three-dimensional forms. Of course, any promising area has to be closely examined on the ground with some of the other tests, but the aerial method is particularly effective in eliminating barren areas which would otherwise have to be examined on foot, in possibly dangerous and disagreeable territory, taking months of valuable time to explore and map.

Once a likely area has been chosen, surface samples are chipped away and then examined under a microscope to determine the rock's composition. The paleontologist, the geochemist and the lithologist all inspect the preliminary samples to determine what fossils, oil traces and minerals are present. If these tests are promising, additional samples will be secured by hand augurs going down ten or twelve feet, enough to give a clue as to the direction of the outcropping and if it indicates an anticlinal structure beneath.

If the examination continues, light rotary rigs will be brought in to secure cores of rock from as deep as two thousand feet, though shallower samples are taken as drilling proceeds downward.

Even with all of this scientific study accomplished, only one in nine wildcat wells produce oil or gas; three of every one hundred find enough to be commercially profitable. Odds are long and expensive.

It was a long time after the oil industry got going that "rockhounds," as geologists were first called, were considered acceptable to any working oil man. Before that, underground substances were sought in many most curious ways.

Need for water in remote, dry areas has been a necessity for man since recorded history. The stars and other portents were used, but in the United States dowsers and doodlebuggers were first used to discover water, then brine, then oil. These "water witchers" were members of an unorganized craft which was composed of those with the "touch" who carried a branch in the shape of an elongated "Y." The dowser held the two ends, one in each hand, and then began walking over suspected areas. When an underground stream was located, the one remaining free end would dip violently towards the ground, indicating the exact spot where the well should be drilled. Water wands cut from a peach tree were said to be most effective.

When seeking some substance other than water, the wands were often "baited." The wand was smeared with oil, or a small container of the substance stuck in the crotch of the dowsing Y. Other dowsers were simpler, using a bottle of oil concealed within a chamois bag and dangled from a string tied to the wand. The string would jerk, like a fishing rod when a trout strikes, when the substance had been located underground. Some dowsers claim that iron bait on wands will work fine for water, but nonferrous materials should be used in seeking oil.

The doodlebuggers or dowsers, as they were called among many other things, scoffed at scientists bent on discrediting their methods. The dowsers said that such efforts showed the scientists' prejudices and skepticism. The scientists, the dowsers claimed, sought their knowledge only from books, not wisdom.

Some of their claims seemed extravagant, but they had competitors. There were those who claimed "x-ray eyes," insisting they could "see" underground oil streams. One man traveled with a small Negro boy who carried a length of stovepipe, padded to fit and protect his face. He also toted a portable stool. At selected spots, he would seat himself and peer into the pipe which had been pressed into the ground. "I don't see nothing," he would usually say. But when he "saw" a river of oil below, there would be a triumphant cry. He had varying success in locating oil, just as everyone else did; but it was a dramatic way to locate it, and wells were enthusiastically dug —at first.

Other apprentice sorcerers claimed they got a pain in the soles of their feet when they walked over oil reservoir rock. Others were "oil trompers," who left deeper than normal footprints when they walked over oil deposits.

Abram James, in 1867, was driving with three companions from Pithole to Titusville. As they passed the William Potter farm, James suddenly began talking in a strange language, one which his friends later identified as the Seneca Indian dialect. Suddenly James jumped from the surrey, ran a short distance and fell to the ground in a trance. Fortunately for the annals of oil folklore, his finger dug into the soil at the exact spot where oil was later found. Seneca Oil, of course. What else could explain the phenomenon? Because a paying well was developed, James, in a similar manner, was successful in locating other paying properties, and he died a rich man. Who, then, could honestly say it was only folklore?

No matter how weird or fantastic the method of obtaining oil, the role of sheer luck afterwards always was an important part. A few feet one way or the other, and Drake's well would have waited for another pioneer.

In the later days of the oil industry, Hal Greer, a prosperous Texas lawyer given to deer hunting and smoking good cigars, which he always lighted with a flaming wax taper to assure an even full ignition, killed a fine buck and dressed it out. He then walked to a nearby pool to wash up and rest a bit. His first thought was a cigar, which he lighted with a taper, flinging it into the pool. To his surprise, the water caught fire, clinging to the small gas bubbles which popped to the surface. Coming from Texas, Greer knew enough about potential oil that, before leaving the area, he purchased three acres surrounding the pool. Later, he leased it to an experienced driller, who went 1,085 feet without a sign of oil. He wanted to quit. Greer convinced him to go down fifteen feet more. A gusher came in which paid them handsome royalties for the rest of their lives.

Probably because of that and George Hook's experience, the drilling of an extra fifteen feet became a superstition in oil lore. George Hook drilled a dry Texas hole. Unable to meet payrolls and other obligations, Hook sold out for debts. Drilling was continued for another fifteen feet, and an entire oilfield of wells was brought in. "Never sell a well until you have drilled fifteen feet deeper," Hook commented later. It became an adage in the oilfields.

Vagrant luck always hovered over the oilfields. The Humble Oil Company drilled a well three thousand feet deep and planned to perforate the casting at that point, since the cores indicated good oil sands. The torpedo somehow stuck seven hundred feet above the planned depth and, to dislodge it, the charge was exploded there. A ten-thousand-barrel well was brought in from an oil strata which had been overlooked as the drilling had probed deeper.

Of course, mislocation was the cause of some successful and many duster wells, according to Samuel Tait in his book *The Wildcatters*. A classic case was that of C. A. Canfield and J. A.

Chanslor, who obtained a lease in the Coalinga district of the California desert during 1895. As they hauled the heavy cumbersome rigs, their wagons became bogged down for the seventh time! Disgusted, they decided to drill where they were stuck. Their rigs were erected, and the results brought in an entire oilfield, the fabulous Coalinga district.

A natural and outspoken partner or follower of the seers is the lease salesman. Most of them are honest; some are shrewd, with just a touch of larceny in their hearts. Others are swindlers. But a lease salesman has to be a fast thinker. He obtains property on which a well might be drilled for, say, $50,000. With that as a basis, the underground farmer—the lease dealer—sets out to sell seventy-five 1 per cent interests for $1,000 each. That accomplished, he will retain 25 per cent control and have a $25,000 cushion whether or not the well comes in! This is a perfectly legal device and is still employed by successful salesmen.

Occasionally, the lease dealer will sell more than 100 per cent of a well. To keep free of the grasp of federal officials because of postal fraud, he must drill a well as promised in his brochure. But if it comes in, the promoter will cap the well rather than try to pay more than 100 per cent of the oil being produced.

The wildcatter, an oilman who drills in unproven oil areas, stands to lose everything if the wells don't come in. The promoter doesn't, because he seldom puts up any money of his own. Even then, he hedges all of his bets.

Naturally, there were many fraudulent schemes, some so flagrant as to be almost a joke to everyone except those who invested in the blue-sky companies. The *Boston Transcript* published a semiburlesque prospectus: *The Munchausen Philosopher's Stone and Gill Creek Grand Consolidated Oil Com-*

pany. It was capitalized for four billion dollars and already had nearly fifty dollars in cash as working capital. Dividends were expected to be paid twice a day, except on Sunday. Officers of the company were listed as S. W. Indle, R. Ascal, D. Faulter, S. Teal, Oil Gammon and John Law.

Shares of the famous Antipodal Petroleum Company were to be issued in $10,000 certificates, which would be sold for 25¢ each to stimulate public interest. The plan for the company was to drill only one well; but this one was to reach China, and therefore draw oil from both sides of the world with only one hole.

Occasionally the con man was hoist by his own petard, with wells appearing where none were really expected. In one such case, Jeff Murray bought 35 acres of California land for $200 per acre. The area, now known as Huntington Beach, is a few miles south of Los Angeles along the Pacific coast. Murray didn't like the barren patch and decided to unload. He purchased sets of encylopedias, and subdivided his property into 420 lots, which made the individual lot cost about $17. Setting the price at $50, Murray threw in a set of books with each purchase as he worked the New England states. The sales campaign was an astounding success, but few of the eastern residents ever inspected their land. And as the books became outdated, they were moved into the attic along with the grant deeds to the California lots. Murray recovered his $7,000 several times over and disappeared.

In time, Standard Oil moved into the Huntington Beach area and hit oil. A frantic search for owners of the Murray "Encyclopedia" parcels was made, but few were found. One contacted—Ezra Hapfield, was offered $300 for his lots. With typical down-east caution, Hapfield became suspicious. He paid up back-taxes and left for California. His was the key lot in what became the Bolsa Chica area, the site of a well which was heard a hundred miles away when 20,000 barrels a day blew into the sky. Hapfield became a rich man because of simple suspicion.

His was an exceptional case, because there were few laws to protect the verdant investor. Any claim, no matter how preposterous, could be published. One advertisement showed a picture of the federal prison at Leavenworth, Kansas, stating: ". . . The doors of this prison will open to receive [name of the promoter] if he fails to make good on every statement made to the public. . . ." It was prophetic. The promoter mentioned was later jailed there for mail fraud.

Land or lease owners were only a trifle less eccentric than the others in such a chancy game. The chemistry of sudden wealth often worked strange miracles on what had been normal people. The landowners universally believed that oilmen were out to cheat them, depriving them of royalties on oil income. Often rumors circulated that oil men had found oil in the area and were keeping the news secret until they could get cheap leases. When a well caught fire, everyone clucked knowingly: "This well got away from them and caught on fire. All they could do was to put it out."

One persnickety landowner demanded an electric log showing progress on the drilling (a complicated device which records rock strata and other vital drilling information). The driller slyly agreed and included this clause in the agreement:

"If a well should be developed, then instead of the lessor being furnished an electric log, he shall have the right to be lowered head first down the casing, equipped with a two-celled flashlight as far as seven thousand feet. . . .

"Lessor will be lowered at a rate any prudent operator would lower his lessor into a well. . . . Should the line part, immediate fishing operations shall be commenced and in no event shall the lessor be cemented in, the well plugged and abandoned. . . ."

Of course, the most scandalous of all lease holders was Coal Oil Johnny, but there were others who matched in a small way his eccentricities.

One successful lessor contented himself with a $40 Stetson and $5 worth of bananas.

A coon hunter who owned a dozen dogs bought all the red meat his dogs could eat, then purchased a touring car, offering to take all of his friends for a ride if they could find room with his dogs.

A blind Negro beggar bought two Cadillacs and each morning had his sons drive him to his favorite corner, where he continued his tin cup business.

The perennial request for a new ax to chop kindling was one of the legends which grew, but a poignant touch was added to the lore when one man, a potential millionaire, said he would get his wife not only a new ax but a grindstone with which she could sharpen it. Better, the stone would be turned with one of the newfangled engines run with gasoline.

Most of these were the odd exceptions. Many, many people invested in blue-chip stocks, bought land or otherwise hedged against the day when a well might become depleted and barren.

Promoters, too, had their failings. One, arriving in an oilfield, couldn't find any accomodations. He walked to the nearest saloon and whispered about a new strike of oil which had been made over the hills, a few miles away. Soon there were ample rooms for the lease salesman, and he spent the night there. But the next day, the excitement became so great that the town began to empty. With the excited chatter of new wells, based on the story he had invented, he decided to leave and see if there was something to the rumor after all.

9

Spindletop

As society changes and progresses, markets are created for new and unusual products. Whenever there is a demand, someone will appear to fill the request.

During the 1890s, Charles E. Duryea built a gasoline-powered "horseless carriage" in Chicopee Falls, Massachusetts. Another inventive young man, Henry Ford, was tinkering with what would become a Model T, and thousands of other auto designs. Those "crazy" Wright brothers were studying the construction of a heavier-than-air flying machine. Everyone knew they should stick to their bicycle business.

All of these machines were designed to use the volatile gasoline which was then a useless by-product of crude oil. Standard Oil was, perhaps, first to offer casinghead gasoline. It was exhibited at the Columbian Exposition at Chicago, along with lantern oil, lubricants, axles and valve oils, ointments and gasoline.

It was the opening of the age of energy.

Spindletop answered the need for oil in unbelievable quantities. Spindletop is probably the best known of all individual wells in the United States. It showed the way for exploration for oil around salt domes, and pointed the direction to the so-called golden crescent along which quantities of oil still lay.

Besides, it was probably the first well drilled by a trained engineer, and it is doubtful that it could have been brought in by anyone less.

Real credit for finding the Spindletop field should go to a man with the unlikely name of Patillo Higgins, who has largely been forgotten. He doggedly sought oil in Texas for ten years before Spindletop blew in at 10:30 A.M. on January 10, 1901.

Higgins had been considered a restless, undirected man by most of his friends in Beaumont, Texas. He worked at one job or another until he lost an arm in a logging accident on the Neches River. That traumatic experience sobered him, and he opened a real estate office in Beaumont, where his father operated a gunsmithing business. The Higginses were regarded as able, stable people, and friends were pleased when Patillo settled down, joined the Baptist church and began teaching a Sunday school class. Instead of spending his idle time in saloons, Higgins studied engineering, chemistry and geology. His studies had been excited by the discovery of a nearby area where pools of water tasted strongly of oil and sulfurous gas vented nearby. Higgins occasionally took his Sunday school class on outings to the area called Spindletop Springs, about four miles south of Beaumont. His youngsters were delighted when Higgins poked a cane into the ground, then lighted the escaping gas. The area also abounded in the eerie phenomenon of Saint Elmo's fire, a ghostly display of dancing night lights often observed as brushlike, fiery jets caused by electricity.

Conducting his real estate business, Higgins had the opportunity to examine most all of the land surrounding the city of Beaumont. He was particularly interested in the Big Hill region, which included Spindletop Springs. Riding the flat land, Higgins spotted an outcropping of what proved to be high-grade brick clay. It was a valuable find, because all brick had to be shipped into Beaumont and locally manufactured building brick would be a great economy. Higgins convinced several Beaumont merchants that a local brickyard would be a profitable investment for all of them. They agreed, and Higgins was sent to inspect eastern brickyards to perfect plans for the Beaumont kilns. Higgins almost forgot his real purpose when he suddenly realized that all of the kilns were fired with natural gas or crude oil. Higgins learned that the fuel was cheap, efficient and burned at a constant temperature.

"Without oil to use in your ovens, there's no use in even considering a brickyard," Higgins was told.

With that in mind, Higgins extended his examination to the Pennsylvania oilfields and some refineries. Not knowing any better, Higgins couldn't understand why, if oil was found in Pennsylvania, it couldn't also be found in Texas. Higgins knew where: Big Hill!

Even before he returned to Beaumont, Higgins wrote the U.S. Geological Survey requesting petroleum geological information, particularly concerning the Gulf Coast areas. Higgins read the U.S.G.S. material, comparing its text with what he had actually seen on Big Hill. Higgins became increasingly excited about Big Hill the more he read. He shrugged off the withering brickyard venture, because Beaumont merchants had become more interested in a furniture factory.

Higgins began boring holes in Big Hill, finding what he believed was evidence of oil no deeper than ten feet. When Higgins saw a small sign: "For Sale Cheap: 1077 acres," he was irreconcilably committed to go ahead. The land could be purchased for $6 per acre, or he could obtain an option for several months for $1,000. Higgins didn't have the necessary money, but was determined to raise it. He drew detailed maps of Big Hill, shading in those areas which seemed to clearly indicate evidence of oil and gas. Everyone was indifferent. He was not able to raise a cent.

One man, particularly impressed with Higgins' conduct since he had returned to Beaumont, was George Washington Carroll, a prosperous lumberman who offered the $1,000, provided Higgins could supply the money for drilling and development work.

Higgins contacted Captain George Washington O'Brien and J. F. Lanier, both of whom owned land on Big Hill and agreed there might be oil. They offered to put their land in a company formed to exploit and control Big Hill. Higgins was elated.

Higgins drew up papers for the Gladys City Oil, Gas and Manufacturing Company. A garish letterhead displayed smoking factories, brick buildings, oil derricks, storage tanks and other dreams. While the company then had little more substance than smoke, Higgins' plans were unbounded: fire stations, hospitals, city hall, railroads and schools—including a college! Higgins might want for money but not for confidence. The first directors' meeting was held on August 10, 1892.

Higgins borowed money to begin work, but it was months before the first rig was hauled up the hill. Despite Higgins' objections, Walter Sharp was hired as driller. His equipment was too light for the 15,000-foot well which Higgins believed would have to be drilled. The rig was flimsy. Higgins was right, but the delay was disastrous. Creditors closed in on Higgins, and he was forced to sell some valuable land to satisfy them.

The first well, a dry hole, discouraged all of the backers except Carroll. The Savage brothers were hired to drill a well on percentage. Dry. They tried a third well and a lease arrangement. A duster.

At that time, the Savages were using the cable tool equipment. Later, using rotary tools, the Savages brought in the magnificent Caddo Field in Louisiana, one of the most important discoveries after Spindletop.

Higgins was discouraged but not defeated. Three dusters shook his theories about oil under Big Hill. He sought the advice of Robert Dumble, the Texas state geologist, who agreed to inspect Big Hill. Dumble couldn't get away and sent an assistant, William Kennedy, instead.

Higgins met Kennedy when he stepped from the T & N O Railroad car and whisked him away. Higgins didn't want anyone to see him and reveal what was going on. Higgins, supremely confident that Kennedy would confirm his oil theories and wanting to surprise his townsmen, didn't want chance disclosing the investigation. Higgins took Kennedy on a Big Hill tour, showing him well logs, samples of rock, accounts of gas pockets and other important data. After a brief inspection, Kennedy stated flatly that no oil would be found on such a coastal plain.

Higgins felt his confidence plummet. Where had he erred? There must be oil on Big Hill. Higgins hustled Kennedy out of town, for he didn't want anyone to know of Kennedy's opinion. But the young geologist wrote an article for the *Beaumont Enterprise* warning away any potential investors "*not to fritter away their dollars in the vain outlook for oil* in the Beaumont area. . . ."

Higgins was near the end. He talked with Carroll about a lease, if he could manage to get one successful well drilled. As a last resort, Higgins decided to advertise in a manufacturing journal, asking for qualified and financially capable people to invest in an oil lease on Big Hill, a gamble which Higgins believed might return millions.

The answer Higgins sought was not long in coming. There

was only one, but that was enough to bring together two of the greatest oil pioneers in American history: Higgins and Captain Anthony F. Lucas.

Captain Lucas was a handsome, six-foot-tall man, broad and erect. He wore a full mustache, and his eyes were blue and soft. Higgins liked him immediately.

Lucas was a qualified engineer, engaged in salt mining and related activities since resigning his commission in the Austrian Navy because of lack of opportunity for advancement. Lucas held views similar to those of Higgins: that oil occurred in relation to salt domes, surface evidence of sulfur, salt, gas and crude oil. The Beaumont situation was similar to areas which Lucas had seen elsewhere.

Lucas was enthusiastic about Big Hill. He secured a lease and option agreement on 663 acres for $31,150, paying $11,050 down and agreeing to pay the balance in two equal installments at 7 per cent interest. Higgins was promised a 10 per cent share, and Carroll matched that with another 10 per cent share. The The Gladys City lease was operative again. It would be simple to expand if the first well was a success.

Lucas brought in a rotary rig, too light for the structure Higgins expected. Working from June to January, they obtained three one-gallon jugs of greenish, heavy crude oil. Even on the brink of success, disaster plagued them. The drill pipe collapsed requiring extensive fishing and redrilling operations. Because the work had gone so slowly, Lucas was nearly out of money. Lucas' wife, who had some money of her own, insisted that work continue. The Captain could have what money was available, and when that was gone, they would sell her jewels, even their furniture and clothes. Backed by her determined confidence, Lucas felt he couldn't fail.

No Beaumont investors were interested in Lucas' three jugs

of oil, so he packed them and went to New York. There he planned to discuss the situation with the Standard Oil. Lucas realized that taking the Standard Oil in was a dangerous gamble, but it had to be done.

The weather was cold when Lucas arrived, and he decided to test his oil by leaving the jugs outside for the night. Next morning there was no sign of thickening, almost a sure sign of high-quality oil, and proving the oil could be stored and transported in cold weather. Calvin Payne from Titusville was sent to inspect Big Hill. Payne and Cullinan, who went with him, were executives in Standard Oil. Both men were considered shrewd oil operators.

"You'll never find oil here," they told Higgins after examining the property.

By this time, Lucas was as committed as Higgins and would brook no opposing opinions. Frenetically, they tried to borrow money.

Two other U.S. Geological Survey scientists sought out Lucas. Dr. C. Willard Hayes and E. W. Parker wanted to personally examine the Big Hill. Their report was also negative. They pointed out that a 3,070-foot well had been drilled in Galveston County, in terrain similar to Big Hill. Not a drop of oil was seen. Besides that, the well cost almost $1 million, a figure far beyond the resources of Lucas and Higgins.

Because of his previous unhappy experience with the U.S.G.S., Higgins hustled the geologists out of town before they, too, could issue an unfavorable report and further damage money raising possibilities.

Help came from an unexpected quarter: Dr. William Battle Phillips, professor of field geology at the University of Texas. For prestigious reasons Phillips didn't want to publicly oppose the opinions of the U.S.G.S., but he firmly believed that Big

Hill contained oil. He gave Lucas a written introduction to John Galey, who was one of the shrewdest men in the oil business. Just then, he and Jim Guffey were developing the Corsicana Field in Texas.

Lucas was entirely honest, relating failures, adverse opinions and other discouragements. Because of the adverse reports, Galey and Guffey were cautious. They finally agreed to advance $300,000, with tough terms. No one, not even Higgins, was to know they were backing the operation. Higgins was to be given 10 per cent from Lucas' share. Higgins, however, retained rights to a thirty-three-acre parcel in the middle of Spindletop Heights, land which would be immensely valuable if a worth-while well was brought in.

Moreover, Lucas was required to obtain lease rights for as much of Big Hill as possible, paying minimal prices. The plan involved 15,000 acres, and Lucas wasn't to draw any salary while three test wells were bored. Because of clouded titles, Lucas wasn't able to obtain all of the leases Galey and Guffey wanted, but the prime land was leased. Galey and Guffey borrowed $300,000 from Andrew Mellon. Al and Curt Hammill were hired to drill the well; Peck Byrd and Henry McLeod were their helpers.

Lucas was away when Galey arrived to stake the well site. Mrs. Lucas took him to Big Hill. Galey saw a hog wallow, full of foul-smelling sulfur water.

"The farmers bring their hogs here to rid them of both fleas and the mange," Mrs. Lucas explained. The wallow was surrounded by a salty, alkaline soil, which was spiked with marsh grass.

Galey studied the landscape a moment and then hammered the stake near the wallow. "Tell your husband to drive the first well here. And tell him that we are going to hit the world's biggest well," Galey said in good humor. "Also, let me know if either of you need anything. Anything at all. I'll see you have it."

Following his instructions, the Hammills spudded in the well, obtaining lumber for the derrick from Carroll, who said it would be free if they brought in a 5,000-barrel well. The Hammills agreed. They had a good contract at $2 a foot for a well which could be put down with rotary equipment for 35¢ a foot. After the derrick was built, they installed a new piece of equipment.

Drill bits in the Corsicana Field had been cooled with clear water. The Hammills improved the system. Water was gathered in a slush pit and a herd of cattle driven through it until it was thick and muddy. This mixture was then forced down the drill pipe and recirculated, cooling bits and sealing walls to keep out impurities.

Lucas' projects seemed star-crossed. Even with the Hammills' excellent rotary equipment, trouble was encountered at 160 feet. Egg-sized gravels there resisted the bits. Failure stared them in the face again.

To continue, the Hammills drove an eight-inch pipe, with a four-inch pipe inside, straight downward with a pile driver. A circulating pump was used to wash up the sand as the pipes inched downward. It took them more than two weeks to penetrate 285 feet of sand and gravel.

Once through that and other gumbo, the bits ran into a gas pocket which sent muddy water and gas shooting halfway up the derrick. The pumps were run 24 hours a day to keep ahead of the muck, which meant that one man worked 18 hours every third day.

Even while the Hammills were sweating out conquest of the gas pocket, gravels and drilling, Lucas was happily obtaining more leases. He never lost faith that a well would be brought in, and they would naturally want to be in control. His confidence was justified when a rock bit was brought up which Lucas identified as limestone, containing traces of sulfur.

"You have hit the salt dome," Lucas said. "We are right."

Al Hammill was working the long shift when the string of tools began to slip easily into the hole. The pumps hummed quietly, and by morning light Hammill could see an oil scum forming in the slush pit, where the mud recirculated. Curt Hammill and Peck Byrd showed up for work. Byrd was sent for Lucas.

Once there, Lucas ordered that more drill pipe be shoved into the hole until the string was resting on hard rock at 880 feet. But 300 feet above, heaving sands had penetrated the shaft. Lucas had the four-inch pipe withdrawn and a larger one inserted. No one was entirely sure just what was happening underground, but these were active signs of oil. The men were exhausted, and it was time to knock off for the Christmas holidays. All of the crew quit, and drilling was shut down.

Work was resumed after New Year's and the bit had gone to 1,020 feet when it stuck in a crevice. Curt Hammill broke the rotary chains trying to free the string. They telegraphed Corsicana for more bits, chains and other equipment, asking that they be put on that night's train so they could be picked up next morning.

That night a huge ball of fire seemed to hang over the top of the derrick. It was Saint Elmo's fire.

"That's a good sign," Lucas told his wife, Carrie. "Sailors believe that ships will come safely to harbor with a flare like that. With the fire hanging over our well, it is a sign of good luck."

Next morning, January 10, 1901, Lucas paid an early visit to the well; then he returned home to eat breakfast. It was a briskly cold morning. Lucas noted that the cloudless sky was bearded with smoke from chimneys throughout Beaumont.

Higgins was on his way to sell some timberland in Hardin County, a sale which would relieve some of his pressing debts. He met Al Hammill on the way to the railroad depot to pick up supplies. They exchanged greetings and parted.

Once the new bit had been screwed on, the drill pipe was lowered, with Curt steering it from the derrick about forty feet above where the drill disappeared into the earth.

At 700 feet the action began. Mud burbled up, and then the well began to roar. Mud continued to spout up in large quantities, higher and higher, until it spouted through the crown block at the top of the derrick. Curt Hammill was covered with mud before he was able to slide to the ground. Al Hammill and Peck Byrd were running away from the rumbling well, each in a different direction.

The roar increased to a crescendo. Then six tons of drill pipe was belched from the well, crashing through the top of the derrick. The pipe broke into sections, the various lengths falling to the earth and sticking like spikes driven in by a giant.

As quickly as it started, the noise and mud flow ceased. The men waited a few minutes, uncertain what to do. When nothing more erupted, they returned to clean up the mess and repair the damage. They described the desolation in terms which would wither a drill pipe.

It appeared that Big Hill had fizzled again.

10

The Gusher

As the Hammills and Byrd shoveled away the mud and debris from the gas blowout, the earth began to shake and an unholy roar issued from the well. It was followed, as before, by a devil's breath of natural gas. The men dropped their tools and ran. Peck Byrd fell into the slush pit and came up looking like a dripping chocolate toy. At a safe distance, the men paused and looked back. From the well came a spout of oil, green and heavy. It shot higher and higher, finally going through the top of the derrick.

"Get Lucas," Hammill yelled at Byrd. The oil was a greenish plume feathering the Texas sky.

Lucas rushed to the well, almost falling from his buggy in his anxiety to get near the well.

"Oil. Oil. Every drop of it," Hammill shouted at Lucas. The noise from the well was deafening.

"Thank God. Thank God," Lucas murmured as he looked into the sky.

The gusher reached 200 feet into the air, and a freshening breeze spraying the oil for miles over the landscape. But there was enough to settle into pools nearby. Lucas and the Hammills began shoveling earthen tanks.

Word was flashed to the world, and by the time Patillo Higgins returned later that day, his friends welcomed him with the greeting of: "Millionaire. You're a millionaire."

Higgins knew that was not so, but he planned steps toward that end.

A special express train was immediately dispatched from Galveston to Beaumont, and every seat was taken. The size of the invasion increased by the hour. Oil leases were hawked at the railroad depot. One lease was offered a reporter for $1,000. He refused. A companion bought it, sight unseen, and sold it for $5,000 within an hour the same way. It brought $20,000 within a few hours. The all-pervading oil madness was rivaled only by the wildest days of the gold rush.

Galey, Cullinan, the Haywood brothers, D. R. Beatty and a hundred other experts rushed to Beaumont. Galey bought newspapers along the way to learn any new developments, and was excited to note that the Lucas Well was estimated to be gushing 30,000 barrels a day. In reality it was 100,000!

"This well marks the opening of the liquid fuel age," Galey told a *Beaumont Enterprise* reporter. Galey tempered his remarks by adding that it would be years before engines, boilers, lamps and stoves could be converted to the use of the new, cheap fuel. "We now have the problem of getting this flow under control. Such waste can't be permitted."

Arriving with Galey was D. R. Beatty, an energetic, imaginative Galveston real estate agent interested in oil properties. He was in the offices of the *Galveston News* when the news flashed about the Lucas Well. He got aboard the specially charted train sponsored by the *News*. Not having time to get any clothes or more money, Beatty arrived with only a $20 bill in his pocket. With the reporters, Beatty lit out for the Lucas Well. Once there, Beatty separated from the others and talked with a man named Lige Adams, who owned about 200 acres a half-mile from the gusher. Beatty sympathized with Adams and walked home with him. Because Adams needed money, Beatty was able to smooth-talk him out of a lease with a deposit, and a promise of one-eighth royalties with the assurance that work would begin within 30 days. Adams finally agreed when Beatty said he would also hire him as camp cook.

Beatty borrowed enough money to get a rotary rig, hired the Sturn brothers to drill and managed to get Bill and Jim on the job on the last day of the option. Just before dawn on March 26, Jim Sturm was blasted off the drilling platform by a gas eruption, followed by a mammoth gusher of oil. That dispelled the comments by some experts, who claimed the Lucas Well, now called Spindletop, was a fluke, a crevice well that was getting the oil from a thimble of oil rock.

With the boom burgeoning, the future looked bright. But it turned bleak when fire flashed through the oil slush pit at Spindletop. Flames roared hundreds of feet into the air, and the heat was intense. Fortunately, a light breeze kept the flames from the well itself. Lucas ordered a back-fire to be touched off on the other side of the pit. Towering walls of flame began to approach each other, and the vacuum created between them sucked up hundreds of gallons of oil, which exploded high into the air from the intense heat below. When the opposing walls of flame met, the explosion rocked Beaumont, and shards of fire fell like a fiery blizzard. But the main fire had been extinguished, and the remainder was quickly controlled. The superstitious insisted the fire was a sign to stop further work on Big Hill.

"Spindletop became the El Dorado of stock company promoters who drilled wells so thickly that there were five hundred on 150 acres," Samuel Tait wrote. "Derrick legs overlapped, and the boards were laid from one to another to permit crews to escape in case of a blowout. . . ."

With the backing of Andrew Mellon's money, Galey and Guffey were putting down two more wells. Scott Heywood arrived from San Francisco, getting a fifteen-acre lease from the Higgins Oil Company. Heywood, an adventurer and rainbow chaser in the goldfields, was one of four vaudevillian brothers, and he obtained his backing from Captain W. C. Tyrell, a merchant at Port Arthur, not far from Beaumont. Heywood chose a site between the Lucas and Beatty wells, hitting oil the first time, and the colorful Heywood Brothers Oil Company was formed.

Higgins brought in a successful well, finally proving his unflagging faith in Big Hill. With only six wells producing, the Texas holes were issuing more oil than all the rest of the world.

Success stories were as common as leases. A Mrs. Sullivan owned a small parcel near Big Hill. She was usually called "Mrs. Slop" because she collected garbage to feed her swine. Even after she sold the lease rights, with royalties extra, for $35,000, Mrs. Slop next day whipped up her team and picked up swill as usual.

Beatty cashed out one of his leases for more than $1 million and was equally successful with others. But no one who invested with Lucas, Higgins or Heywoods made big money. Lucas might have if he had permitted the use of his name in connection with a fraudulent oil promotion. He refused $1 million.

Higgins made some money by suing both Lucas and Carroll to obtain the 10 per cent rights he had been promised. He asked for $6 million but settled out of court for an undisclosed sum.

These suits and the enormous amounts paid for lease rights focused world attention on the vast amount of money involved in the oilfields. An estimated $200 million was invested in the area, which was given the name "Swindletop" because much money was obtained fraudulently.

There was some serendipity from Spindletop. It broke the Standard Oil monopoly. Previously, Standard Oil controlled virtually all oil production from Pennsylvania, West Virginia, Indiana, Ohio, New York and California; but when the Spindletop Field came in, the oil produced there was more than that from 37,000 eastern wells, six times the California production and, with five other wells to come in soon, eventually more oil than in the rest of world.

Except for the difference in the texture of the treasure being wrested from the earth, there was little difference between the oil and gold boom camps which flourished earlier in American history.

Carroll, an extremely pious man, would permit no saloons, gambling or brothels on any Gladys City Company land. But they blossomed nearby. The first was a section called the Log Cabin. Single men and women lived in that area, men with families in Gladys City and riffraff in a section called South Africa. Order was maintained by Sheriff Ras Landry, a typical slow-talking western officer who favored white boots, hat and a fast gun.

There was need for supervision. Fifty thousand people roamed Beaumont's streets, many carrying sheaves of $1,000 bills ready to buy leases on the spot. "Hello Boomer" was the universal greeting, and no one thought of money per se—only of oil and how to get it.

Many of the frantic buyers and sellers didn't know anything about the oil industry; but that didn't matter. The frenzied atmosphere made many astute investors lose their perspective. Too many wells were drilled, releasing too much of the vital gas pressure which forced the oil to the surface, and worse, when the oil was obtained, it was for a waning market. Autos were not yet in mass production. Other engines and boilers hadn't been converted. Even so, many major oil companies got their start in the Beaumont field: Gulf, Texaco and Sun were but a few.

The Texas Company sprang into life because "Buckskin Joe" Cullinan, formerly operating in the Corsicana area with Standard, rushed to Spindletop when it gushed in, along with John "Bet-a-Million" Gates, who had extensive investments in Port Arthur and interests in both American Steel and Wire and the Kansas City and Southern Railroad, which served the Texas area.

Guffey was planning a pipeline to Port Arthur, where he would build docks and refineries to handle the oil from Spindletop. Cullinan had similar ideas of "buying it in Texas and selling it up north." The Standard Oil wanted the Texas oil too; everyone wanted it because of the unstable market. A syndicate, Hogg-Swayne, merged with Cullinan in an amicable arrangement; the syndicate supplied the money and Cullinan the know-how. Others moved in, and Cullinan began to buy land, pipe, storage and dock space. The syndicate spent $600,000 before the first income was received. Subsidiary companies were formed, in which Gates was a heavy investor. He was never permitted to get control because of his profligate gambling tendencies.

Gulf Oil came on strong from the time of its birth among the derricks of Spindletop and remains a major contender among world oil producers, refiners and marketers.

The backers of Gulf Oil were the Mellon brothers, William, Andrew and Richard, who had also backed Galey and Guffey. The Mellons were no strangers to the oil business, even before Spindletop. They stubbornly fought Standard Oil for an interest in Pennsylvania and other oilfields. When Standard Oil blocked oil shipments through control of railroads, the Mellons laid a 271-mile pipeline to a waterfront at Marcus Point, not far from Philadelphia. Rockefeller then realized they could jeopardize his entire empire and bought them out for a $2.5 million profit for the Mellons. The money put them in an excellent position to further compete with Standard Oil, if necessary.

Galey and Guffey bought the Lucas Well with Mellon money, a success which excited all investors. The J. M. Guffey Petroleum Company was formed as a vehicle to finance expansion. Wells needed to be drilled and refineries built; tugs, barges, pipelines and a hundred other things were needed at once. The Mellons were heavily committed, and the Andrew Carnegie interests provided necessary money for planned expansions. With such blue-chip backing, price of the stock skyrocketed when it was formally chartered on May 16, 1901. A second company was needed for marketing and refining outlets, and Guffey wanted the name of Texas to be included. But he found that it had been pre-empted by Gates and his partners in Texaco.

"Balked on the word 'Texas' but determined to adopt a name that would identify the company with Texas oil," Craig Thompson recounted in *Since Spindletop,* "company officials borrowed the word 'gulf' from the nearby Gulf of Mexico. Thus, but for the difference of a few weeks, all that is known by the name of Gulf might have been called Texaco by some other name—possibly Gulf."

Spindletop demonstrated to the world what gigantic oil production meant, but when Texaco drilled in the nearby Sour Lake area (about 25 miles from Beaumont), it seemed the

Edwin L. Drake (in top hat) drilled America's first commercial oil well in Pennsylvania in 1859. Shown with Drake is Peter Wilson, who assisted him in the venture. *Courtesy Drake Well Museum*

A street scene in Oil City, Pennsylvania.
Courtesy New York Public Library

Transporting barrels of oil down Oil Creek.
Courtesy New York Public Library

Pithole City, a Pennsylvania oil boomtown.
Courtesy Shell Oil Co.

Fire raged out of control at many early oil wells.
Courtesy New York Public Library

A Pennsylvania petroleum field in the 1860s.
Courtesy New York Public Library

Samuel M. Kier sold bottles of crude oil in 1847 as a remedy for numerous ailments.
Courtesy Drake Well Museum

An early form of tank wagon used for delivering kerosene. *Courtesy Gulf Oil Corp.*

Kerosene and motor oil used on a Texas farm.
Courtesy Farm Security Administration

Drilling for oil in Pennsylvania in the 1870s.
Courtesy New York Public Library

A cluster of oil derricks at Spindletop.
Courtesy Gulf Oil Corp.

An offshore drilling rig in Cook Inlet, Alaska.
Courtesy Union Oil Co. of California

world would be drowned in oil. The price for crude oil dropped to 5¢ a barrel!

Overproduction was not the entire reason. Texas oil was heavy and black, saturated with sulfur and based on asphalt. This was in dire contrast with the Pennsylvania oils, considered "sweet" because they were light, greenish and paraffin-based—differences which made it easy to refine into "sweet-burning kerosine."

Guffey was unconcerned. He hired excellent chemists to seek new products and improve refining. Some wondered if it had been worth the gamble when, in August, 1902, Spindletop sputtered, hesitated and then ceased to flow.

Utter disaster faced the Spindletop investors. The Mellons tried to unload on Standard Oil, but that company had had enough of Texas oil. An affiliate had been fined and banished for violating state antitrust laws, and Standard Oil withdrew entirely.

"We are out," an official said. "After the way in which Mr. Rockefeller had been treated, he'll never put another dime in Texas."

The Mellons said afterwards that Standard officials were amused with the situation in which Gulf Oil was entangled. Not only had it suffered a total Spindletop loss, but that was compounded by the fact that a long-term contract bound Guffey to deliver 4.5 million barrels during each of the 20 years at 25¢ a barrel to the Shell Trade & Transport Company. With the collapse of Spindletop there was no cheap oil, and eastern oil sold at 30¢ a barrel or more. Fulfillment of the contract meant disaster, but Shell Company officials, whose specialty was trade in mother-of-pearl, renegotiated the contract and Gulf Oil survived.

Wildcatting operations were launched, new aggressive men

were hired to develop oil markets and pipelines were built to Sour Lake, Batson, Humble and other fields. All this, combined with hard-knuckled, shrewd management, enabled Gulf Refining Company to finally emerge as the enormous Gulf Oil Corporation, which absorbed its subsidiaries on January 30, 1907.

Fires continued to ravish the area, some lasting as long as a week before being stifled with steam and sand. One, started by a lightning stroke, was so intense that rain condensed from the clouds snuffed out the flames. But even with the fires, Big Hill still looked like a toothpick container, with the image of all the derricks lifting skyward. Eventually the derricks were disposed of and replaced with simple surface pumping units.

Though Spindletop seemed exhausted, Lucas was recognized as the pioneer developer and given medals and testimonials. Higgins was almost entirely ignored, despite the fact that his observations had made salt dome oil discoveries one of the major tenets of oil geology. The oilmen knew the effects of these domes, in pushing up strata, beds of shale, limestone and sandstone, was much like laying a handkerchief flat on a table, then picking it up by the center. The domes are forced up through tectonic, internal earth movements, it is believed.

But few of the boomers cared about such details when they dealt in leases involving thousands of dollars with no more concern than buying a bottle of whisky, which often proved to be a better investment than the leases themselves.

The Crosby House, which included rooms, saloons, a café and even two art galleries, was the center of activity. Men stood on chairs to shout bids on leases. Land was traded for cash only. Only the largest operators could issue a check for a lease. Many promoters never saw the land on which leases were bought, sold and exchanged so rapidly. They were content to skim off their paper profits.

Land was at such a premium that the flower garden which bordered the Crosby House was subdivided into six-foot-square cubicles. Cardboard and two-by-fours were used to make partitions between 'offices"; packing cases and barrels were used for desks and chairs to consummate oil lease deals.

Saloons stood ready to accelerate flagging activity, and they remained open 24 hours a day. They were cited every Monday for breaking the Sabbath laws against serving liquor. A $50 fine was of no consequence when men stood six deep at the bar on Sundays. And when successive wells blew in, the entire city of Beaumont drank itself dry. For a time, half of all the whisky consumed in Texas went down Beaumont throats.

Whisky was better than the water, which was not fit to drink unless one wanted to hazard the ravages of the Beaumonts, a particularly virulent diarrhea. Even doctors warned against the water. Boiled water was sold on the street for 5¢ a cup, which was more than oil cost. But the local chapter of the WCTU (Woman's Christian Temperance Union) was so dismayed with the situation that they arranged to supply free water at various points in the teeming, sweating city.

Spindletop stories were often enlarged and embroidered, but the truth was actually enough to convince a cynic that an immense amount of treasure was flowing from Big Hill. Employees of the *Beaumont News* believed what they read in their own sheet and indulged in lease trading. The newsmen's venture netted them $30,000 profit—reason enough for a monumental celebration which lasted two weeks, until the money was all spent. Meanwhile, the newspaper had to suspend publication until printers and editors regained their sanity and returned to their linotypes, composing stones, headaches, hang-overs and wrathful wives.

As production of Spindletop Field seemingly faltered, fake promotions became even more blatant and insistent. Responsible oilmen, such as Higgins, Lucas and Cullinan, repeatedly warned the public against the frauds. Even so, there were over 500 Texas oil companies and many more in other states. Printing companies couldn't fill the massive orders for stock certificates. Men were hired to do nothing except sign the certificates with fictitious titles. Many worked on a piecework basis, getting so much for each signature.

One of the wisest investments was that of Colonel E. M. House. House purchased all of the rights in the streets and alleys of Gladys City for $90,000. This gridiron gave him a site adjacent to all producing wells, and the $1 million stock issuance sold in less than 36 hours. House's company paid some handsome dividends until legal difficulties finished off the venture.

Even though many of the Beaumonters got rich in the boom, most of the residents were disgusted with the continual hoopla, crime and fraud which arrived with the Spindletop gusher. Everything was crowded, and rain made the streets a morass so deep that locomotives were rented to pull wagons from the mud. Prices for every thing were out of reach of the average man.

In time, an imposing monument was erected to mark the site of the first gusher in the United States—the one that showed the way to the age of energy, the age of liquid fuels. On the memorial, tribute was paid to Captain Lucas. Patillo Higgins, whose dedicated faith brought it all about, was not mentioned.

11

Osage Oil

No one paid much attention to the U.S. Indian Agent in 1853 when he reported the presence of extensive oil springs in the Chickasaw and Choctaw Nations, many of them along the Wichita River. The report said the oil was highly regarded as a remedy for various diseases and was sought by Indians and others. The federal officials paid no attention; nothing was done.

A small oil discovery in what is now Kansas, about twenty years later, caused no comment either, even though the commercial use of oil as axle grease had been demonstrated by farmers in the area.

Robert M. Darden took an interest in the reports and decided to do something about them. He arranged a joint meeting with the tribal chiefs of the Chickasaw and Choctaw Nations, reaching an agreement to drill on their lands on a 50 per cent royalty arrangement. Darden formed the Chickasaw Oil Company in 1872 and prepared to work. But the government stepped in and refused permission to drill. Darden went broke without discovering anything.

Others followed Darden with uniform lack of success. One of these was Michael Cudahy, a scion of the meat-packing family, who obtained leases for all of the Creek Nation in 1884. In 1894 his well at Muskogee showed some faint traces of oil at 1,120 feet. It was probed to 1,800 feet and then abandoned.

Although there hadn't been any encouragement, wildcatters continued to makes leases and drill in Indian lands. Almost exactly thirty years after the discovery of the Drake Well, Edward Byrd found oil along Spencer Creek, which ran through the Cherokee Nation. Byrd's well brought in a half-barrel a day at thirty-six feet. It wasn't commercially successful, but it gave hope to Byrd and those who followed his lead.

Byrd pushed his advantage with Chief Bushyhead, who leased him 94,000 acres of Cherokee lands. Byrd's second well produced three barrels a day at seventy-five feet. The next well was a duster. Byrd's fourth brought in five barrels at about 100 feet. These wells were completed in November, 1899, and are considered the start of commercial oil production in Oklahoma, though the first issue of oil was merely a rill compared with what was to come.

Unfortunately, the Department of Interior begun to tangle the drilling tools with red tape. Byrd, making only a marginal living anyway, wouldn't put up with this obstacle and left the region. George B. Keeler, who operated a trading post along the Caney River, proposed to stick it out. Keeler's determination was shored by the fact that he had ridden over most of the area and remembered many places where his horses refused to drink because of foul smells or oil scum. Oil was there, and Keeler believed he could bring it up.

Agreeing with Keeler's opinions were two other Bartlesville pioneers, Bill Johnstone and F. N. Overlees. Together they leased 200,000 acres of Cherokee land. With the leases in their possession, they asked Galey and Guffey to contract for the drilling; but after several delays in agreements, they became disgusted and left. By that time the options had expired, and nothing was done.

Keeler then asked Cudahy to undertake the drilling. But

with his own experiences and examples of others, Cudahy wouldn't move until the Secretary of Interior Ethan Allan Hitchcock approved leases. Most leases were disallowed, except for one square mile which included the township of Bartlesville.

After immense difficulty in moving rigs (one 65-mile section took three weeks), a derrick was erected near the Big Caney bridge at Bartlesville. On April 15, 1897, the well blew in, spouting 50 to 75 barrels a day.

Bartlesville grew slowly. The village had been founded by Jake Bartles, who built a flour mill on the Caney River in 1877. Keeler and Johnstone moved in to start a general store, and the tiny trading post attracted considerable business from traders, cattlemen and Indians. Practically all of the business transacted there was done by barter. Store goods, brought in from Coffeyville, Kansas, were exchanged for furs. It was a brisk, profitable business.

But oil was not amenable to such a system. Oil had to be packed somehow, because there were no storage tanks, pipelines or other adequate transportation to markets. Bartles surveyed a railroad right of way to intersect the Santa Fe in Kansas. He sold his maps to the railroad, and the first engines chuffed into the area during 1899. Behind the engines were tank cars. The problem was solved.

Bartlesville flourished as other wells were made, but it was overshadowed by later fields. It was the start of the powerful Phillips Petroleum Company and the lesser-known Indian Illuminating Oil Company, founded by Frank Phillips and H. V. Foster, two of the richest and most powerful men to spring from Bartlesville wells.

Despite the faith of the original locaters, an immediate boom didn't develop. Cattle continued to be the principal industry, but oil excitement increased when the Territorial Governor of Oklahoma wrote the Secretary of Interior: "At Pawhuska . . . two tests have produced petroleum of superior grade. . . . There are many indications of oil at other places in the [Osage] reservation. . . ."

The Territorial Governor continued to file optimistic reports with the Department of Interior, pointing out that wildcatters were locating and bringing in successful wells of high quality oil.

". . . There is unmistakable evidence of both oil and gas at many points within the Territory. In Payne and Pawnee counties are several springs where the water is polluted with oil, and in the Chickasaw Nation and the Kiowa and Commanche reservations are other oil springs. . . . Paying wells are being put down in all parts of the Creek country. . . . A local company is drilling at Guthrie. . . ."

Because of the specter of continuing federal meddling and the barren, miserable areas in which oil was located, there was no frenzied rush to Bartlesville. Some of the reluctance was dissolved by Dr. J. C. W. Bland, a physician married to an Indian woman from Tulsa, and Dr. Fred Clinton. They bought an oil rig which had been shuttled off at the Red Fork siding for the express charges due against it and began to drill.

Four months after Spindletop gushed, the Glenn Pool was tapped. The Sue A. Bland Well, named after the doctor's wife, was spudded on May 10, 1901, and gushed in with 600 barrels on June 24, the first such well in Oklahoma Territory.

With a bottle of oil taken from the well clutched in his hand, Dr. Clinton hurried off to visit Dr. Bland, who was hospitalized because of an appendectomy. Clinton obtained a power of attorney from Mrs. Bland to assure the Indian rights. Only then they looked, tasted and savored the oil in the small bottle. It was too good to be true. But it *was* true. Still, there was much to do.

Without even changing his oil stained clothes, Dr. Clinton took the first available train to Muskogee, where the federal Dawes Commission ruled over the Five Civilized Tribes as a paternal body.

Though it was after midnight when Clinton arrived, he woke Dr. F. B. Fite, a man known to have influence with the Dawes Commission. Together they tested the oil on shavings and then in a lamp. Fite was impressed and arranged for a meeting early the next morning. Clinton was granted lease rights on the forty acres surrounding the well site because of Mrs. Bland's tribal rights. Clinton returned to Red Fork, desperately tired but victorious and happy.

Despite the fast-flowing gusher and the electrifying news of another oil discovery, the boom had defects. Oil was selling for only $1 per barrel, but transportation, barrels and production costs took 90¢, which didn't leave much for the leasor. Land sold briskly because the area was now free of federal control. Bland and Clinton refused $40,000 for their well and continued to purchase "deeds of possession" whenever available. But many leases were acquired by men who didn't have ample funds to continue exploitation, and the boom collapsed. By this time, Tulsa and other oilmen had been bitten deeply by the oil bug and they liked the feel.

Wariness about future moves by the federal government, deflation and the fact that only a few could reach key men in the Interior Department hit the area hard. Some "moonlight" wells brought in small production. The demand for oil was still limited, but there was hope ahead.

R. E. Olds built 425 Oldsmobiles in 1901. Henry Ford produced about a thousand cars in 1903. Standard Oil pointed the way in demonstrating how enormous amounts of money could be made by efficiency, foresight, courage and ruthlessness.

Bob Galbreath was about to pack up and leave when Bob Glenn asked him to look over an oil-stained limestone ledge on his farm, which he owned with his Indian wife. Glenn pointed out several crevices in which oil glistened—greenish and translucent. Galbreath was impressed with what he saw and convinced Frank Chelsey to join with him in a test well using the Glenn (Indian) allotment lands, about ten miles from the nearest producing wells. It was strictly wildcatting.

This was strictly a "poor boy" well, and every possible economy was effected. The well was not even cased. But they brought in a gusher on November 22, 1905. Because of the remoteness of the Glenn farm, the drillers were able to keep their success a secret until ample backing was obtained and additional leases obtained. Then they cased the first well: Ida Glenn #1.

With the added financial backing Galbreath drilled a second well, which was a duster. But the third brought in 700 barrels a day, and the next, 2,000 barrels.

With these leases proven up, the big operators began to move into the Glenn Pool area and obtain leases. Galbreath would not sell to any Standard Oil representative seeking leases on land he controlled. To pressure him into changing his mind, Standard Oil cut off transportation, steel storage tanks and markets. Glenn fought them successfully, even though he had to store the surplus oil in earthen tanks and take other emergency measures. Galbreath was successful enough that, even when the field began to decline, he sold out for nearly a half-million dollars. He and his associates had been offered $1.5 million at one time.

Galbreath sold to J. E. Crosbie, a Canadian who acted for the Gypsy Oil Company, owned by the Mellon interests. Crosbie was an experienced, skilled oilman and was given substantial percentages of the leases he obtained for Gypsy Oil. Crosbie did his job well, getting good leases and laying a pipeline to move the oil to market.

Crosbie did so well that his income was rumored to be several thousand dollars a day. There were many other millionaires who emerged from the Glenn Pool operations, but few were more successful than Crosbie.

Large-scale operations took place when Ed and Henry Foster arrived in the Indian Nations, seeking investments to enhance their already substantial fortune. Another brother, Barclay, remained at Westerly, Rhode Island, to manage the family-owned bank.

Through John Florer, an Indian trader, the Fosters obtained leases on all of the Osage nation lands on March 16, 1896. The money from the leases would be divided among all tribal members. The lease provided for a 10 per cent royalty and a $50 lease fee for each well for the next ten years. Though it was signed by Chief Bigheart and later formally approved by the Department of Interior, it was considered a scandal because of the immensity and exclusivity of the blanket lease being given to one group. Though the first few wells were dusters, and two of the Fosters died during the first year of operation, the company, Phoenix Oil Company, was merged with the Osage Oil and the Indian Territory Illumination Oil companies. The leases were immensely valuable. Early-day investors, even creditors, were usually paid in stock shares rather than cash, and all received many times the amount of their original claims. Production increased enormously, perhaps thirtyfold.

Long before there was any thought of oil or any other interest in the region, the Osages had roamed an enormous area, including most of Oklahoma, Kansas, Arkansas and Missouri. But with the encroachment of the whites, and the unilateral treaties or agreements which were dishonored, the Osages were squeezed into a smaller and smaller area, each poorer than the last. Their lands were seized at the convenience of the government and sold to settlers, usually for $1.25 an acre. Unknowingly, the federal government was setting the stage for an ironic situation which made the Osages the richest Indians in the world.

By 1925, every Osage Indian received at least $13,200 from oil royalties annually. Each Osage had been given "head right" —that is, one full share of the total amount of money received by tribal chiefs in oil rights, leases and royalties. These head rights could be willed to heirs but could not be sold, mortgaged or otherwise encumbered. As a consequence, one survivor of a large Osage family was receiving $112,000 annually through the head-right inheritances.

The irony began with the leases sold to the Foster brothers, which covered an estimated $1.5 million for ten years. The federal government later withdrew about half of the lease, but wells on the remaining acreage kept production to the tens of thousands of barrels a day.

In 1898, the Curtis Act spelled out the treatment of the Five Civilized Tribes: Seminole, Creeks, Cherokees, Chickashaws and Choctaws. The Osage Nation was apart from this arrangement, and they were allowed full control of their lands for the benefit of the entire tribe. The others retained surface rights, but none of the underground mineral or oil rights. These were to be husbanded by the federal government, supposedly in the best interests of the tribes.

Retention of full oil rights by the Osages made a great deal of difference in which lands were developed. Oilmen were understandably wary of the government, which might step in and revoke or alter a lease without warning or regard for the investors. Because the Osages made independent leases with the approval of the Secretary of the Interior, their land interested the oilmen first. Even hard-nose Galey and Guffey moved out of the Cherokee Nation when the government tried to interfere.

When the first Osage leases expired, a financial fight erupted between the oilmen; but the issuance of leases was given over to the federal government by virtue of a rider on an appropriation bill. The good faith of the Interior Department was always suspect, but the prize was too large to ignore. There were 680,000 acres of Osage land open to lease. Since all of this was proven oil land, a bonus lease payment of at least $10 per acre, plus a one-sixth royalty, was considered minimum. That meant $6.8 million in bonuses alone.

Two U.S. Senators, Chauncey Depew and Harry New, were approached by representatives of the Prairie Oil & Gas Company, a subsidiary of the Standard Oil. The blandishments they made are not known, but they cornered about one-half of all available land and paid only one-eighth royalty instead of one-sixth.

It was a mighty coup, and Governor Charles N. Haskell was infuriated. He wrote a scorching letter to President Theodore Roosevelt, excoriating him as chief trustee for the Indians for allowing such blatant favoritism. (This incident may have shaped T.R.'s later conduct against trusts. There is no evidence that he knew anything about the chicanery, if indeed it existed at all.)

The whole system smelled of a conspiracy. The Department of Interior was swamped with requests for lease approvals, but only those backed by the biggest companies got any attention. While the leaseholder was limited to 4,800 acres by any person or company, clerks in the Department, some making no more than $10 weekly, were able to successfully bid in leases and claim responsibility for $5,000 in the bank or other financial stability to fully develop a lease!

It was something of an irony, too, that land allotments made by the Dawes Commission, before the Glenn Pool came in, were to Indians for agricultural purposes. The best possible lands were given to those with proven enterprise and inclination to work. But the more indolent Osages didn't ask for land, and when their land allotments were assigned they were given the hard-scrabble, barren sections. These areas were to prove up as the most valuable of all the oil lands. Crazy Snake and thirty others had to be put down forcibly because they didn't want to be bothered with any land allotments at all. It was against tribal custom, Crazy Snake insisted.

Unequal, uneven treatment of the Osages amounted to a federal crime. Indians were allowed to fritter away their wealth, while others became enormously wealthy. Incompetents, and 625 Osages were declared such, were assigned guardians who often lined their pockets while their charges lived in poverty.

John Palmer, a Sioux who had been adopted into the Osage tribe, convinced federal officials that a census of the Osages must be taken. When it was completed, 2,229 Osages were counted. Each was assigned a head right, or 1/2229 entitlement to all tribal income. Naturally, as the original Osages died off, the income accumulated into a few hands, a situation which led to some startling events.

12

Oil Makes A State

It wasn't until the enormous oil reservoirs in the Osage nation had been discovered that anyone seriously considered that the Oklahoma Territory might become a state. But with its sudden affluence, the "land of the red men" became a state on November 16, 1907, with Charles Haskell as first governor.

Some hoped the dignity of statehood would end the wild frontier days, but those optimists had reckoned without the oil boom cities. There was much sin among the derricks.

Tulsa claimed to be the oil capital of the world, and it was, no doubt, though it also boasted some primitive living conditions for such an exalted title. One man converted a stable into a hotel, of which he said: "It took six months to get the horseflies out so residents could enjoy themselves. . . ."

Oil production was now estimated at 43.5 million barrels, with Prairie Oil & Gas the largest buyer. Prairie built a six-hundred-mile pipeline from Oklahoma to Indiana refineries. The Texas Pipe Line Company and the Gulf Pipe Line Company laid lines to the gulf ports and refineries there.

The profligate manner in which the Indians threw away their oil fortunes had many counterparts in some of the big spenders who couldn't stand the prosperity of a newfound oil fortune. One was Billy Roesser, who was only thirty-two years old when he sold eighty oil leases for $350,000. He kept other leases, which were estimated to be worth $1 million.

Roesser bought the finest house in Tulsa, gave away the furniture and had the house redecorated by a St. Louis expert who installed $16,000 worth of "real genuine oil paintings." The grounds were landscaped with imported plants which cost $15,000. He ordered a car and had it shipped express from the factory, which cost almost as much as the $6,000 car. It wasn't long before Roesser needed ready cash—even more than he needed his newfound friends, who were always badgering him for loans. When his wells were dusters or stopped production, Roesser had to sell everything he had. When the debacle ended, his estate was worth only $40.

"There's only one way to go and that's up," Billy Roesser confidently told his wife. "All we had was $40 when we started, so we haven't lost any ground."

Roesser heard of a likely oil area not far from the Glenn Pool and hired a car to go there. With nothing more than an oral agreement, Roesser leased a tract from a farmer and immediately sold these rights for $600. Not far away, Roesser obtained another lease, got a well put down on a percentage basis and then sold out for $20,000. It would be nice to relate that Roesser parlayed his stake into millions. But it wasn't so. Billy Roesser lost all of his money in the next venture and was never able to recover financially.

Joshua Cosden was another wildcatter—one who succeeded. Even though he made and lost millions, Cosden persevered to become a philanthropist, an asset to Tulsa and to the oil industry generally. Cosden was a reporter on an eastern newspaper when he somehow learned of a new process to refine high-grade gasoline. It was extremely valuable because of the swelling demand for gasoline with the rapidly expanding automobile industry. Cosden purchased a lease on a small well near Bigheart in the Osage Nation and built a small refinery some distance away. The crude was hauled in an old wagon which leaked so badly that Cosden followed it with a bucket to recover what he could. His refinery was blown apart, and it burned twice. It actually produced very little, but Cosden never lost faith and eventually overcame his obstacles.

The oil business was chancy, hazardous and heartbreaking. It required the confidence of a burglar and the faith of a zealot. As a consequence, the oil industry has more than its share of rage-to-riches stories about the men who braved the odds.

". . . The entire American oil industry is but the lengthened shadow of the independent oil man, whose form and substance are stamped indelibly over its whole structure," Wallace Pratt, a famed oil geologist, observed.

Frank Phillips was a young man with the following stated ambition: "I knew a barber who always wore a pair of spring-bottom striped pants. I made up my mind that I had to become rich enough so that I could afford to wear striped pants, even on weekends. . . ."

When he was fourteen, Phillips went to work in a barber shop. Within ten years, he owned all of the barber shops in the town. Phillips realized that a plateau had been attained, so he sold cigars, tobacco products and homemade hair oil to fill the needs of his customers and enhance his income. Phillips sold bonds for the local bank. Before he was through, he owned the bank and had married the banker's daughter.

In his selling trips, Phillips visited Bartlesville and saw another opportunity there. He opened the Citizen's Bank & Trust Company, and it was only a short step from there to deal actively in leases and the financing of oil developments. Frank sent for his brother, L. E. Phillips, to help stack up the money his bank was making. Before long, two other brothers were brought into the business, which finally became the Phillips

Petroleum Company, still one of the giants in the oil industry. Phillips, before he died, was able to buy all of the spring-bottom pants he wanted—even for weekends.

Harry F. Sinclair, who founded the Sinclair Oil Company, is alleged to have secured his first stake to buy oil leases after collecting insurance money from a gun accident which removed his big toe. He was a capable and intelligent young man, anxious and determined to succeed. Like Frank Phillips, he obtained his goal by seizing every opportunity and shaking the last ounce of profit from it.

Sinclair was a clerk in his father's drug store in Independence, Kansas. When oil was discovered nearby, young Sinclair began dabbling in lease rights, making quick profits and further exciting his ambition to get rich. With sufficient working capital, Sinclair moved quickly into the Glenn Pool area, which was just then awakening. He secured leases before prices went beyond the limits of his resources, but he then found himself competing with the giants: Gulf, Standard, Texas and others. Still, he wasn't dismayed.

Sinclair circulated rumors about how much more he would pay than the others, and thereby got a free chance to negotiate with all of the leaseholders. And when one Cherokee, Frank Tanner, was reluctant, Sinclair bought a local baseball team so the Indian could be signed on as pitcher (which was Tanner's sole ambition). Sinclair got the lease.

Sinclair's career continued on its meteoric orbit until he foundered in the scandal of Teapot Dome, a black chapter in the history of the oil industry and the federal government. Essentially the matter concerned the leasing of naval oil reserves and alleged attempts to defraud the government.

One of the men with whom Sinclair was associated while dealing in leases was "Mad Tom" Slick, perhaps one of the greatest gamblers in the oil industry. Slick was a man of the times—when oil deals involving thousands of dollars were sealed with no more than a handshake and a promise. It is claimed that for years after Slick died, his executors paid off agreements which were substantiated by nothing more than scribblings on the back of an envelope.

Slick had grown up in the Pennsylvania oilfields, working among the derricks while a youngster. He knew all there was to be learned about oil and seemed to have an intuitive sense for locating oil. Slick wasn't infallible. He missed on some of his wildcatting operations. As a consequence, he spent about as much time looking for new partners to finance a well as he did seeking out new oil sites.

Slick was described as a gaunt, cadaverous-looking man whose deep-set, brooding eyes were accentuated by a crown of white hair. He arrived in Oklahoma following moderately successful leasing operations in the Robinson field in Illinois, and the discoveries in Kansas. Slick was in on the Glenn Pool, but his interest was drawn to another area, about forty miles west of Tulsa. Slick had the feeling that the scrub brush-covered hills and ravines contained oil even though the first two wells drilled there had been dusters.

Slick, seeking an oil lease, appeared hungry and tired at the log cabin of Frank Wheeler, located near the Cimmaron River in Payne County, asking for overnight lodgings. There wasn't much room to spare. Wheeler lived in the cabin with his wife and nine children, eight of them girls. Though the Wheelers were usually hostile toward strangers and unbidden guests, they allowed Slick to stay with them.

Wheeler had never been able to make much of a living from his 160 rock-encrusted acres. He pieced out the family income by hiring out as a mason, when he could find work. Even with the moonlighted income, there was a little to share with strangers. Besides, oil prospectors were always wandering through the area promising much and producing little. Anyway, there wasn't an active well within five miles.

Slick was leasing land with the backing of the Jones brothers, who owned a bank in Bristow, Oklahoma. Slick obtained a lease on the entire Wheeler property. It wasn't until Slick had also leased much of the surrounding territory, more than a year later, that he put down the first well, which was a duster at 2,000 feet. The Joneses withdrew their backing.

Slick knew something the Joneses didn't. The sands hadn't produced any oil, but Slick believed he could smell it. Slick studied the rock cores, mapped the geology as best he could and then choose another site about three miles away. This was along a creek, perhaps chosen because of the (discredited) "creekology" theory. Despite Slick's persuasive ways, the Joneses refused to advance any more money. They had bought all the suitcase rock they wanted—more than enough.

Slick had met and resolved similar cases of timidity, and he wasn't to be turned away this time. He invited the merchants of Cushing, a nearby village, to a public meeting, pointed out how the entire area would prosper if he was able to fully develop his leases on the Wheeler farm. Slick explained the geology of the region as best he knew it, following with an impassioned plea for a total of $8,000 to finance drilling. He was certain of success, and promised one-half of all future leases to the investors. They would be worth millions when the first well came in. Unfortunately, many residents of Cushing had been "slicked" before and lost their money. They weren't to be taken in again. Slick's generous offer was flatly rejected. Not one cent.

For once, Slick was at wit's end for financing. Then he remembered C. B. Shaffer, a wealthy Chicago speculator and a friend in the Pennsylvania oil fields. Slick borrowed $100 for fare to Chicago, where he met Shaffer and was able to convince him to finance a Cushing well. Shaffer not only agreed to put up the necessary $8,000, but gave Slick enough extra money to lease another thousand acres in the Wheeler farm area. Drilling started immediately.

It was March 10, 1912, when Slick's first well roared to life as a gusher. Thousands of barrels of oil spouted into the air, and the surrounding area quickly became a lake of oil. After strenuous efforts, Slick's drilling crew were able to cap the well and they began covering up all evidence of oil to keep the secret of the gusher while Slick obtained more leases.

Slick had one advantage. There were virtually no roads in the remote area and only a few cars and buggies to drive the roads which did exist. The poverty-ridden farmers had no cars and no telephones.

By the time Shaffer and J. K. Gano, a high-pressure lease buyer, arrived, Slick had nearly throttled any effective competition. He had rented all of the cars in the area, along with the buggies and wagons owned by residents. All vehicles were hauled to an open pasture and put under guard of riflemen. New arrivals could explore the area only on foot. They even had to walk to the first Wheeler well. Slick had done his job well.

Unfortunately, he overlooked one available automobile when he had been cornering the transportation. The owners of that one car got $25 from each person they hauled anywhere, and higher prices for special trips.

Even so, Slick, Shaffer, Gano, the Jones brothers and many others made fortunes overnight. The Cushing Field was big enough to accommodate a hundred millionaires.

Frank Wheeler, who had given Slick the first leases at a time he was so poor he would have had to finance a shovel, was now getting $1,200 a day, an income which slowly shrank to a comfortable $300 a day. Wheeler had spent too many years being poor to fritter away his money. The income went into banks or blue-chip investments. Wheeler bought a comfortable home in Stillwater, and then bought farms as wedding presents for each of his daughters.

Cushing swelled with 6,000 oil-maddened men, most of whom had to sleep in tents or out in the open. There were no accommodations. Because of the antiquated water system and inadequate sewage and sanitation, disease flourished. And with no police authority, crime was more common than good conduct. Money meant little, and life even less.

One man argued over his place in line for a haircut and was shot to death for his trouble.

A whisky drummer, twice arrested for illegal activities, warned that anyone who bothered him would be shot, and made good on his boast when the sheriff tried to arrest him. Although the drummer shot the officer, he was never prosecuted or even jailed.

While there were ample hazards inherent in oil field regions, a new one developed at Cushing and the adjacent Drumright Field. Two automobiles, the "horseless carriages" of those days, were moving in the vicinity of active wells when they both suddenly exploded and burst into flames. The engineers couldn't fathom what had happened. Finally one solved the mystery. Natural gas, allowed to escape from the wells, had seeped into the dips of the roads. Sparks from the exhausts had ignited the volatile mixtures as the cars entered these pockets.

Even tragedy was considered good publicity for the oilfield. While Slick bought the first leases for $1 an acre, publicity forced up prices. Harry Sinclair was offering $8,000 for forty-acre leases, or $200 an acre. More was offered if the property seemed worth it.

"The prices being paid for the property are greatly in excess of what the property is really worth," said an editorial in the *Oil & Gas Journal.*

The observation appeared sound when the 23,000-barrel-a-day production declined. Cushing began to assume aspects of a ghost town. There were now, of course, the know-it-alls who said they had believed it to be a shallow field. But there were just as many oilmen who were sure there was a lot of undiscovered oil. Wells were pushed deeper, and their faith was rewarded. The oil-soaked sands—the so-called Bartlesville Sands—were reached, and they squirted oil by the thousands of barrels a day high into the air. Initial 5,000 barrel-a-day wells were frequent, and drilling became so frenzied that there were not enough drillers to put down the wells. Frantic calls were made to Pennsylvania, Indiana, Texas, Kansas, Illinois and even California to lure drillers, tool dressers, roughnecks or even "boll weevils," as the rookies in the oil industry were called.

Oilmen responded by the thousands, and like numbers swarmed in to get rich. The nearby Drumright Field, named after the Aaron Drumright lease, sprang from nowhere. Cushing, at least, had been a village before the rush. Drumright was a farm with a single house. As the rush got under way, 6,000 people were receiving mail at Drumright. The post office was located in a jerry-built pool hall, and the mail was dumped on one of the back pool tables. Anyone seeking mail pawed through the pile of letters and parcels. As the oilmen shuffled the mail, the letters quickly became so smudged and oil-stained that the last, remaining letters couldn't be identified from the outside.

That Drumright, Tiger Creek and Cushing were all occupied by hard-working, hard-drinking and hard-fighting men was shown by a classic schedule which some anonymous roughneck scribbled on an oil-smudged envelope:

11 a.m.	Get up
11 to 11:30	Sober up
11:30 to 12	Eat
Noon to midnite	Work like hell
Midnite to 3 a.m.	Get drunk
3 to 3:30 a.m.	Beat hell out of them whats got it coming
3:30 a.m.	Go to bed.

Many of these Oklahoma boom towns equaled anything the great gold rushes ever produced. Contemporaries recalled that a person could look in any direction and see a fight or two in progress, with only the participants interested in the outcome.

There were some shootings, knifings and similar brutalities committed, but several murders remained secret until some of the storage tanks were dismantled and the oily bones of numerous men were found at the bottom of the containers.

As a result of these disappearances, there were frequent advertisements in the newspapers or oil journals seeking the whereabouts of missing husbands or sons, who were last seen in the Cushing or Drumright areas. Some of these men, no doubt, vanished into the storage tanks.

Such continued rowdyism interfered with the orderly production of oil, and there was an organized attempt to suppress the illegal whisky traffic. Prohibition had been one of the requirements imposed on Oklahoma for its admission as a state. The restriction was designed to help the Indians.

Four U.S. marshals were sent in to put down the bootleg trade, but selling liquor, just as in later prohibition days, became an underground secret. This arrangement satisfied everyone. The men had just as much to drink, except that it couldn't be done publicly. It was thought to improve oilfield efficiency because there were fewer fights and similar outbursts. Sales of lemon and ginger extract and a foul mixture called "choc" (named after some unknown, unlamented Choctaw) reached heroic proportions.

John Galey visited the boom camps when they reached their peaks. With Guffey, Galey had once controlled all of the land which comprised both the Glenn and Cushing Fields! Together, or even singly, they were worth astronomical amounts. They had let their options drop because of interference by the Interior Department. But Galey was not a whiner.

". . . When I consider what might have happened if we had held on to all of our holdings, it staggers me to think how rich we would have been," Galey was reported to have said in a book titled *The Greatest Gamblers.* "But I have no regrets. It is the fortune of the game and I am glad to see so many of the producers making good and am glad to know that the percentages of dry holes is so small. . . ."

To the end of his life, which concluded a few years after his visit to Oklahoma, Galey continued to search for oil, never regretting the failures or oversights. It was the search, not always the realization, that kept the wildcatters happy.

13

Seminole & Bowlegs

Enormous amounts of money flowed through Oklahoma from the early oil discoveries, and the flow has continued into the present. The Seminole received very little of it, largely because of the inattention of federal officials toward their protection and welfare following the 1906 Indian census. Half-breeds were allowed to handle their own affairs, free of any control whatever. Most of these Indians forfeited, sold or otherwise disposed of their lands, receiving nothing when oil was later discovered. Many of those who retained their lands and received oil income frittered the money away, just as many of the whites had done before them.

Because of their head-right arrangement, the Osages generally fared better than any others. Auctions for the leasing of their lands were held publicly at Pawhuska (the original capital of the Osages), beginning on November 11, 1912.

Colonel E. E. Walters, official auctioneer for the Department of Interior, was a man of commanding appearance. Though his voice was strident, it was also persuasive. When these open-air sales were first held, Walters got only $39,000 in bonus money for the leasing of 107,000 acres, or about 37¢ an acre. A decade later, Walters sold the lease rights to 35,000 acres and received $10 million. This bonus was topping for the Osage cake because they were already drawing one-sixth royalties from many producing wells. As some of the Osages died and the head rights were concentrated in a few hands, the income became immense.

With her income, who was there to deny Lucinda Pittman, an Indian woman who had come into much oil money, the right to choose the color of her Cadillac? The possession of a Cadillac was the sign of affluence to the Indians. Lucinda went to a paint store and picked out the gaudiest color she could find, something between chartreuse and puce; then she insisted that her new car be upholstered and painted the same shade. The auto dealer hesitated, but not for long, for Lucinda offered cash for the car and the extra service. Not long afterwards, rich Indians owned expensive cars in aquamarine, baby pink or magenta. Lucinda had set the style.

Such profligacy was condoned by federal officials, largely because there was nothing that could be done about it. None of the Osages could touch any of the tribal capital, and since it seemed to be everlasting, it didn't make a lot of difference to anyone where the daily income went.

One Osage stalled his car in a ditch and took out a bull whip to stir some life back into it. A passing farmer called: "You'll never bring that one to life. It's dead."

The Osage looked at the expensive car again, shrugged and asked for a ride into town. He signed over the title of the $6,000 custom-built car to the farmer and bought another—one that was alive.

The Osages were a colorful part of the life in the Oklahoma fields and often attended the public auctions, blanket-wrapped and stolid. They were in strange contrast to others in the audience: the Boston dowagers, gentlemen dressed to the nines or roughly clad, redolent of crude oil. All were equally anxious to purchase choice parcels. Colonel Walters was able to deal successfully with all. Walters and his assistant, who checked maps for locations, moved the bidding along briskly, usually in advances of $500 or more. And no one was startled when an opening offer on a particularly desirable tract started at one or two hundred thousand dollars.

Bidders had their own special signs to indicate a bid as the auction bidding became more spirited. Fingers moved, an arm bent, a head nodded—any of these movements might mean thousands of dollars. All bidders knew the land intimately, and knew what their minimum and maximum bids would be before the rap of the opening gavel. Stirring contests, occasionally involving millions of dollars, developed when the major oil companies competed. In one historic day of auctioneering, Walters brought $10 million into the Osage tribal treasury.

Such a flood tide of money led to strange doings. When no other new cars were available in the area, an Osage bought a long, sleek black hearse. The rear door was removed and the Indian had a rocking chair throne installed so that he could look and see where he had been as a chauffeur drove about the Territory. With such riches abounding in an undisciplined society, crime followed. A series of murders resulted.

The sanguinary drama began when Anna Brown, a rich Osage woman, was murdered. Miss Brown had been found shot through the head in a patch of grass not far from Fairfax. Since the Osages were known to be a trifle careless about shooting one another, especially when they had consumed too much bootleg whisky, there was scant official concern about Anna Brown's death.

A few weeks after her demise, a cousin, Henry Roan, was found done in, a bullet in his head. He was seated at the wheel of his new car. Roan had inherited the head right and other valuables from the deceased Anna Brown. Officers also discovered that Miss Brown herself had inherited head rights from Lizzie Q. Brown, an Osage crone who had died under mysterious circumstances. At least, she died without medical care, and there was no subsequent official investigation.

Charles Whitehorn, closely related to all of the other victims, was found dead in Pawhuska, another murder victim.

George Bigheart, William K. Hale and Ernest Burkhart were all mysteriously stricken ill and rushed to the Oklahoma City hospital.

Bigheart, a direct descendant of the hereditary tribal chief, was interviewed by his lawyer, Charles Vaughn, at the hospital just before the Osage died. Though it was presumed that Bigheart disclosed some information about his strange illness, it was never revealed because Vaughn was killed the next day, near Pawhuska.

A pattern was emerging from all of the apparently interrelated deaths. The wealthiest members of the Osage tribe were being systematically killed off. The motive was the head rights, which would be concentrated in a few hands. Terror permeated the Osage tribe. No one could tell who the next victim would be because of closely bound marriages to relatives. Some hired guards; some moved away or took other precautions to protect themselves. Even so, more Osages were slain.

Death and destruction continued. A dynamite blast rocked the neighborhood around the W. E. Smith home. Smith, his wife and a servant girl were all killed in the explosion. Investigation showed that Smith's wife had recently inherited the Brown and Roan head rights. With the death of these three, there was a public outcry for protection and thorough investigation to locate the killers.

Suspicion turned toward Ernest Burkhart and his uncle W. K. Hale, a man of considerable influence in Osage Country.

They had been taken to the hospital with the deceased Bigheart, and appeared to have been out of the way when the latest killings had been accomplished. But there was a nagging suspicion about Burkhart because he had inherited not only incomes (through his Osage wife, Mollie Brown) from some of the slain Indians, but also $25,000 insurance money when Roan was murdered. Burkhart explained that the insurance had been partial security for a loan which had been made to Roan.

Because of their influence, no police officer was anxious to arrest either Burkhart or Hale unless they could gather incontrovertible evidence. The pair were too powerful. The investigation continued.

Bert Lawson, a felon who had been questioned in connection with several other murders, was hauled in again for questioning about the Smith dynamiting. Lawson was easily able to establish an alibi. At the time of all the murders, he had been in jail for one charge or another. Officers were still suspicious. It was too pat. Lawson was taken to Oklahoma City for further questioning. Once there, he suddenly confessed, producing one of the weirdest documents on record.

Lawson confessed to touching off the dynamite charge under the Smith residence, and he then confessed to several other of the murders. Just as Lawson said, he had been in jail and the records proved it, but through connivance with the jailer, he had been secretly released to commit the murders. With the deed accomplished, Lawson returned to jail and was safe from suspicion. Lawson implicated Hale, Burkhart and the jailer, M. A. Boyd, in the plot.

Intensive security was provided for Lawson because of his importance in prosecuting the others, and the whispered threats upon his life. Even though a total of seventeen murders had been committed, Oklahoma officials didn't prosecute. Federal agents were called in. Some of the people murdered were wards of the government, which was excuse enough to prompt intervention by the Department of Justice.

A federal investigation was launched, but most witnesses were afraid to talk. Hale and Burkhart stood in the background. In time, however, Hale, Burkhart and John Ramsey, a farmer, were indicted.

Burkhart was the first to be tried, with the trial set to take place in Pawhuska. Burkhart was arrogant and uncooperative. His blanket-wrapped wife, Mollie Brown, sat impassively in the courtroom watching, though little understanding the proceedings. She arrived each day in a gaudily painted Cadillac driven by a chauffeur.

"I don't work. I married an Osage," Burkhart snapped when asked about his means of livelihood. As evidence of his guilt mounted, Burkhart's arrogance crumbled and he finally changed his plea to guilty. He was immediately sentenced to life imprisonment. But before he left court, he named Hale as the mastermind behind the entire scheme and implicated two other men. (These men had been murdered before the trial opened, apparently to prevent their turning state's evidence.)

After the two lengthy trials, the first resulting in a hung jury after fifty hours of deliberation, both Hale and Ramsey were convicted and given life sentences. The reign of terror in the oil region was over, and one of the strangest chapters in the history of the booming industry ended.

The Seminole Indians didn't fare well with the advent of oil, even though a tremendously rich field was brought in on what had been Seminole tribal lands. Only the more stable Seminoles, who retained their land, farmed it or lived on it, realized any benefits from the black gold which later gushed up all around them. Following the Indian Census of 1906, dwindling attention was paid to the welfare of the Seminoles, who had been forcibly settled on the barren land after two wars in the Florida Everglades against the entire United States. (The Seminoles were never beaten, and they never surrendered.) Those few who had acumen enough to retain their lands reaped a handsome harvest following the discoveries at Seminole and Bowlegs.

Both the Glenn and Drumright fields had peaked before the first discoveries were made at Seminole (which had been nothing more than a flag stop on the Rock Island Railroad route). The first wells at Seminole were drilled successfully by Henry Doherty's Cities Service & Gas Company at the suggestion of Charles Gould, a consulting geologist.

O. D. Strother and Bob Garland figured too, but they remained independent of the others. Strother, a shoe salesman, was an amateur geologist and had an abiding faith in the oil-producing abilities of the Seminole area. Whatever spare money he and Strother accumulated they invested in the purchase of land in the area, or lease arrangements which would not expire until the land had been tested. Bob Garland was a wildcatter who had lived all of his life in the oilfield regions. He operated on the theory: "Fill the earth with enough holes and you'll find oil sooner or later."

Both men, though tackling the problem from opposite directions, proved to be right in their respective beliefs. Unfortunately, Strother died before he could realize his dreams.

The rush to Seminole was immediate and dramatic, a replay of other speculative migrations. In some ways, Seminole was perhaps more brutal and more poignant than others. But the prize was certainly worth the seeking. In time, Seminole,

Bowlegs and adjacent areas were producing 10 per cent of all American oil.

"All of the majesty, the fury, the danger and the urgency of a great oil boom is at Seminole," the *Tulsa World* reported. "All its confusion and strife and litter; its laughter, its suffering, its curses and its victories. And all of its raw materials. . . .

"It had taken skill, patience, perserverance, the overcoming of almost insurmountable obstacles, the hardest manual labor, and millions of dollars to bring in the great Seminole field. And in pride, perspiration and tears the oil fraternity had done it. . . ."

It was a monumental, costly task. Wells, producing or dry, cost about $60,000 to reach the oil sand levels, a price which was about twenty times that of Cushing wells. But with proven wells in sight, there were always men who would finance explorations. Cable rigs were discarded in favor of the faster, more expensive rotary rigs. Most of the wells were drilled to 3,000 feet or more, and some derricks sported signs stating they were prepared to hit either "hell or China," a favorite oilman boast.

One driller scoffed: "We passed hell last week."

Expenses mounted because of the difficulties in moving both men and equipment, hampered by primitive roads—when there were any paths at all. Heavy rains turned the clogged streets into hog wallows with ruts. Cars were often abandoned in the roadways, and it frequently took a day, even a week, to remove the obstacles and to get traffic moving. It took twelve hours to travel the five miles between Seminole and Bowlegs.

Seminole was another Drumright. It had been nothing but an old railroad parlor car which was used as a depot when the trains stopped, and that wasn't often. The only passengers were the Indians or transient oil prospectors who went to Seminole to flag down a train and get out of the barren area. Both Seminole and Bowlegs were little more than tent cities, though more substantial buildings were constructed as soon as lumber could be shipped in.

Liquor was illegal, but no one seemed to know it—or care, if they did. The gambling, dance halls and brothels in Bishop's Alley were reminiscent of the wildest days in San Francisco's Barbary Coast. While these made the most public noise and were the subject of talk and writing, law enforcement officers, aided by responsible citizens, kept the vilest section confined to those who wished to inhabit it. Several murders were committed. Most of the guilty were apprehended and convicted, though they were given relatively light sentences.

One slaying for which no one was found guilty was of a loud-mouthed braggart who had been cut to ribbons in a saloon brawl. Bleeding so profusely that his boots sloshed with blood, he continued to stride about the streets telling everyone how tough he was without seeking medical aid. He continued the public harangue right up to the moment he collapsed and died, having bled to death while he continued to talk.

Drilling in the Seminole and Bowlegs fields continued at a frantic reckless rate. So much oil was produced, tens of thousands of barrels a day, that the market was depressed with the oil deluge. The larger operators got together to form an association which would control oil production and thereby keep the market price firm. The plan was wise but was impossible to enforce because of the independence of the oilmen. The wildcatters had to keep the oil flowing to repay their backers and recapture drilling and exploratory costs. The plan failed, but the attempt at oil conservation, a natural resource, was not lost on governmental agencies. Conservation practices prevail in all modern fields.

One driller frantically drove his bits deeper and deeper, finally reaching the 2,000-foot level, where his tools became stuck. Another driller, a few hundred yards away, encountered the same difficulty. What had they struck? Both noticed bits of iron filings coming to the surface with the drilling mud, and they finally realized that both bits had drifted, cutting crooked holes, and had locked far underground. It was an almost impossible feat, but it happened. Both of the wells had to be abandoned because of the carelessness. Because the days when sensitive instruments ensured the accuracy of shafts, it was only the driller's skill and intuition which kept the bits driving straight down. It was natural that drillers received premium pay.

The Seminole and subsidiary fields were opened. In time, great discoveries were made at Ponca City and Oklahoma City, and throughout the Territory. There were great gushers, great fires and great times. But things were happening farther west, in Texas and California.

14

The West Erupts

Until the discovery of Spindletop, which fully opened the age of energy, oil exploration had been primitive and desultory. The industrial lethargy was especially true in California, which since 1866 had done little to expand its oil production.

Josiah Stanford resorted to boring thirty-one tunnels into the flanks of Sulphur Mountain, holes which produced from one to twenty barrels a day. Even with this minuscule production, Stanford was, for a time, California's leading oil producer. His adits were dug at an angle which would permit the gravity flow of oil into tanks.

Even Professor Silliman's enthusiastic report didn't evoke much interest.

Samuel F. Peckham, writing in a 10th Census report, said: "Early in 1864 this region [between Santa Barbara and Los Angeles] was visited by an eminent eastern chemist, who was so far misled by false representation and by gross deceptions practiced upon him as to induce him to make a report upon this as a petroleum producing region of great richness. This report and others of similar character led to the formation of mining companies representing stock in the millions of dollars, all of which, it is needless to add, was lost to the bonafide investors. . . ."

Peckham was quite clearly referring to Silliman and was warning away any investors considering California oil properties. In the light of subsequent events, Silliman's report stands out as a minor miracle of prophecy and accuracy.

C. C. Mentry, a bearded, heavy-set man, arrived in California from Titusville in 1873. He was a skilled, experienced oilman, a derrick builder, driller and superintendent. Mentry didn't pay attention to the repeated warnings that California oil wasn't worth looking for. Leases were cheap and relatively easy to obtain. Mentry bought one near the Colonel E. D. Baker well in Pico Canyon, where he was withdrawing six barrels a day from only 90 feet.

Mentry punched a well with cable tools and struck oil at thirty feet; but it produced only two barrels a day. He then went deeper—to seventy-five feet—and tripled the output of oil. He spudded in another well across the canyon and brought in a small well. Heartened by these two successes, Mentry hurried to Los Angeles in an effort to obtain leases on the twenty-five acres which lay between the two wells. With the leases in hand, Mentry then sank a well on a direct line between the two wells. A duster. That might have been the end of Mentry's exploration, because he was short of cash—but for the arrival of D. G. Scofield from Titusville.

Scofield, a stern-visaged man who wore a long handlebar mustache, had become a successful kerosine dealer since arriving in California in 1870. He became increasingly interested in the producing end of the oil business after hearing numerous optimistic reports. Scofield knew that San Francisco was the center of financing and oil marketing, but all of the major prospects for wells seemed to be in faraway southern California. Scofield undertook an extended inspection tour of all producing wells. He listened with equal attention to facts and rumors. Of all the leases he saw, Scofield liked Mentry's the best. In the firm belief that he could obtain necessary leases,

Scofield sent for W. E. Youle, a famous Pennsylvania driller. Scofield formed the California Star Oil Company to purchase Mentry's leases and operate two refineries. Mentry was retained as superintendent, and when the new steam-drilling rotary rig arrived, Pico #4 was drilled. Oil was struck at six hundred feet, yielding one hundred fifty barrels a day. It was the best well in California and the first truly commercial well there.

With oil now available in worth-while quantities, J. A. Scott built a simple refinery near Newhall, situated along the Southern Pacific Railroad tracks, which linked both ends of the state. Both production and marketing problems were solved. Newhall was only seven miles from Pico Canyon. Oil was hauled in wooden barrels by teams to the refinery. Finished products went out by rail. Later a pipeline, considered the first west of the Mississippi River, was laid between the wells and refinery.

Success of the Pico Canyon efforts spurred E. A. Edwards to put down a well in Adams Canyon on the land of the former Mission San Buenaventura. A three-barrel well was made. This, added to seepages in the area, was considered enough for construction of another refinery. The Los Angeles Oil Company was organized to develop an 800-acre lease in the Little Sespe region, also in Ventura County.

Scofield continued to scour California for oil, becoming interested in the Moody Gulch strata in the Santa Cruz Mountains, not far from San Jose. Success there would have the advantage of a location within fifty miles of San Francisco. Before Scofield moved, R. C. McPherson, owner of the San Francisco Oil Company, obtained leases on most of the land which Scofield wanted.

McPherson, with a Colonel Boyer, owner of the San Jose Gas Works, teamed to exploit the Moody Gulch area; but the men were never able to get along. Boyer, whose temper was volatile as gasoline, grabbed up a pistol and forced McPherson to sign over the best leases. Boyer then persuaded Youle to drill some wells, after Scofield temporarily released him from their contract.

". . . I ordered lumber and rig timbers sawed at a mill in the Santa Cruz Mountains some miles away," Youle recalled later. "I purchased boilers, engine and tools that were stored in San Francisco—a secondhand outfit but in very good condition. The largest bit was eight inches; the rig had . . . drilled a deep water well. I purchased several sizes of pipe. . . . I located a 1,500-foot coil of drilling cable and sandline; they were Manila. . . . It was certainly some job to make the forgers understand what oil tool work required. However, they tried their hardest and I was satisfied with the bits. . . ."

Youle drove a well to seven hundred feet, where he struck a twenty-barrel well. A second well brought thirty barrels at 1,700 feet. Though it looked like a minor bonanza in oil, it was ephemeral. Wells were sunk in almost innumerable California locations. At McKittrick, the oil was too viscous for use. All Glendora wells were dusters. The Little Sespe leases, however, were producing enough for Edwards' refinery at Ventura to be able to ship kerosine to both San Francisco and Los Angeles markets.

Because the pistol-point policies of Boyer proved too much for even the rough oilmen, suits were filed to legally recover the purloined leases. Men sued for back wages, and Youle attached the entire operation to protect his rig and tools. Moody Gulch work ceased.

Scofield interested Charles Felton, a state senator, and Lloyd Tevis, a banker, in both the Pico Canyon and Moody Gulch properties. Other small companies were merged, and the

Pacific Coast Oil Company was formed on September 9, 1878, capitalized for $1 million. It was the giant in the western oil industry then.

"They could spend a million dollars each and bear their losses if it proved to be a failure, and they were the kind of men who would not blame their advisers for any loss made," Youle observed.

Youle's observation proved to be prophetic. In 1879, the company obtained about 20,000 barrels and sold them for only $40,000, which was scarcely worth the effort. Many of the wells issued oil which was so asphaltic that it couldn't be successfully refined with methods known at that time. These wells were capped, for no one wanted to spend the money to try to solve the problems of refining the complex crude oils.

Heavy-duty, steam-powered rigs finally arrived from the east in 1880. Several wells were spudded in the Pico and nearby Wiley canyons. Plans to drill in Moody Gulch were formulated, and more men were recruited from eastern fields. Daniel Dull, one of the top drillers, came, along with Lyman Stewart, who was to make oil history on his own.

Dull and Youle agreed to work together, sharing contracts, equipment and, hopefully, profits. Youle brought in a 100-barrel well in Moody Gulch, and followed that with five more successful wells, all with about the same production. Scofield ordered fifty railroad tank cars to haul the oil; then he began building a 500-barrel capacity refinery at Alameda Point on San Francisco Bay. Dreams of selling to foreign markets loomed large.

But their oil production was puny in comparison with the $20,030,761 in gold that was dug in California during that year.

Panic gripped the owners of the Pacific Coast Oil Company when the Moody Gulch wells suddenly ceased to flow. Commitments to deliver both crude and refined oil had to be met. There were expenses to pay and a refinery which was standing idle. Drilling rigs were rushed into the Pico and Wiley canyons.

"Get us oil and get it fast," Youle and Dull were told.

The drillers did as they were told, bringing in well after well, steadily swelling production until it reached 500 barrels a day.

Though the causes for the Moody Gulch cessation were not known, the drillers took no chances in putting down the holes, despite the urgent need for oil. They worked carefully and used large quantities of drilling mud to ensure against caving in or the influx of impurities. To give the mud more body, lengths of manila rope were forced into the well shaft. Their care paid off, and the wells continued to produce.

But another serious problem faced them. The Southern Pacific (SP) Railroad, usually referred to as "the Octopus," began charging prohibitive rates to carry oil from the wells to the Almeda refinery. (The Scott refinery at Newhall had closed down, and there were no others nearer.)

Pacific Coast's request for rate reduction was coldly refused by the SP. Undaunted, Pacific Coast ordered miles of pipe and began laying a line to Ventura, where the oil could be shipped to San Francisco aboard makeshift tankers. When the SP saw the specter of defeat and loss of revenue, they "rechecked their figures" and found that oil could now be carried for about half of the previous rate. Pipeline and tanker construction was stopped and brisk oil production continued. Lyman Stewart would be the first to make major use of both pipelines and tankers.

Drillers new to California were warned that rock structures were "something different" from those encountered in other fields. The strata was twisted, faulted, harder and more difficult to drill. Lyman Stewart was ready to meet the challenge. After inspecting some of the Pico Canyon sites, Stewart telegraphed his former partner, W. L. Hardison, asking him to join in the California venture. Having complete trust in Stewart, Hardison replied that he was settling his affairs and sending two complete drilling rigs, and enough drillers, tool dressers and men to operate them.

Stewart leased acreage on Christian Hill and began drilling in 1883, when the new rigs arrived. This well was cemented in and was probably the first to be so treated in California. At least, Youle said it was the first he had ever seen. Cementing was done at the cellar and within the casing to strengthen the hole.

Stewart's first hole went to 1,850 feet, with no oil. Rock corings indicated that they were drilling through twisted, faulted rock which was very hard. The bits hit rock which deflected them to make a crooked hole, and when a new start was made, the tools were lost and couldn't be fished out. The well was abandoned.

The next three wells proved equally difficult and dry.

"The law of averages will take care of us," Stewart and Hardison agreed, even though their original capital of $135,000 was becoming dangerously anemic. They decided to go for broke. The Christian Hill leases were forfeited and new leases obtained on the Smith Farm, along Tar Creek. To reduce expenses, Stewart asked his men to reduce their wages temporarily, but they refused. He shrugged and ordered work to proceed.

Smith Farm #1 was spudded in July 20, 1883 and the first 119 feet cased with pipe and cement. Bad luck plagued them. They hit oil strata at 672 feet; at 1,300 feet oil flowed in small amounts, but it was mixed with sulfur water. Stewart insisted that they go deeper. At 1,520 feet the shaft caved in, their tools were lost and drilling halted.

Stewart and Hardison were too heavily committed to withdraw. A seventh well was drilled. It was a duster. It was an incredible string of failures, considering that these men were experts in the work.

Hardison, who had been bankrolling these ventures, had been a banker in Kansas. To finance costs, he had been kiting checks or overdrawing on his account until the bank was nearly broke. There was a sudden run on the bank, the other depositors wanting their money before Hardison gobbled it all up.

Stewart went to a friend in the California Star Oil Company, explaining that they had been beaten to their knees by repeated dusters and wanted a chance to drill in a proven oil region.

"The Star lease [part of California Star] looks like it might be a winner," Stewart was told. "Why don't you try it?"

Star #1 was spudded in, with Stewart aware that this was the last chance. Success! At 1,620 feet they made a 75-barrel well, taking out more oil in one day than they had from all of the other wells.

"We were elated," Stewart said. "We had finally struck oil, but that was about all. Now we didn't have enough money left to properly develop the oil in Star #1."

It was a cruel irony that success had broken them. Stewart and Hardison tried to obtain other leases but were refused. They finally had to sell out to the California Oil for an undisclosed sum.

Even though it seemed they had been cruelly put upon by fate, it was the Star leases which failed. When Star #1 was drilled deeper, production dropped one-half, and all of the other wells in the area were dusters. The Star lease rights were later proved to be on the northern edge of the Pico Canyon fault traps. But Stewart learned one more lesson. He should have bought, not leased, land where the wells were to be drilled. Stewart and Hardison were determined never to be beholden to anyone again. These men were to be counted among the "greatest gamblers" in the oil business. They made courage and faith do what cash did for others.

With their remaining funds, Stewart and Hardison bought acreage in the Santa Clara Valley, which resembled the Ventura Valley. Sites were obtained in Adams and Wheeler canyons and the Saltmarsh area, near Santa Paula, which was then only a small trading center.

Again, their efforts were star-crossed. A small but productive well was made in Adams Canyon. A second well drilled nearby drained all of the oil from the first. A third drained the oil from the other two. Worse, the oil was "dead" and heavy, with only a small kerosine potential. Half was asphalt, usable to treat roofing paper, coat cast-iron pipes or surface roads, and there was some gasoline content which could be used in certain cleaning fluids. Despite continued defeats, Stewart and Hardison were determined to drill.

Just then Youle hit oil at 1,600 feet near Puente in southern California, a discovery which implied that oil reservoirs might be deeper than previously suspected. The folds, anticlines, faults and other possible oil traps could be mapped by extensive and careful examination of surface rock outcroppings. If it proved accurate, such maps would be a vital tool in choosing oil drilling sites.

"Go into Adams Canyon and see how deep drilling will pay off," Stewart told field superintendent John Irwin. "Perhaps this way we will reach new oil sands."

The deeper drilling paid off. Successful wells were made

which produced from five to three hundred barrels a day. With the wealth of oil now available, they came up with an idea which was a turning point in the oil industry. Boilers, which articulated their steam rigs, burned coal which cost $30 a ton.

"Why not use the crude oil from our wells?" Stewart asked. The fact that it hadn't been done didn't faze him. Raw crude oil clogged burners which were adjusted to kerosine. Oil slowly dripped onto rocks didn't work. Blowers which sprayed oil in fine droplets worked satisfactorily. All rigs were gradually converted, and Stewart became the prophet and crusader for oil as fuel. The market immediately doubled.

Still, they were paying $1 a barrel to transport oil from southern California to San Francisco refineries and markets. The Hardison & Stewart Company, with the Pacific Coast Oil Company, decided to make some changes. Working together, in 1888 they laid a pipeline to connect the wells with Ventura, where two newly built tankers—wooden ships, the *W. L. Hardison* and the *George Loomis*—could carry 6,500 barrels in their holds to San Francisco. SP reduced its rate to 50¢ a barrel, but the railroad had waited too long this time. The oilmen were determined to cut the arms from "the Octopus," and they completed their project.

Oil prospectors roamed California looking for additional oil-rich areas, even though the oil markets were still limited. It would be another decade before extensive use of oil became commonplace. The Coalinga and the Sunset-McKittrick fields in central California were opened. Two great companies were formed during 1890: Union Oil and Standard Oil of California, which agreed to buy all crude oil of the Pacific Coast. Everyone believed there were literally oceans of oil to be recovered in California—even more if other coastline drilling also proved successful.

Until Union Oil drilled the Adams 28, a gusher, interest in extensive exploration was still moderate, even though oil was being found in impressive quantities. Some of it was so heavy with asphalt that the wells had to be heated with steam before the crude oil could be lifted to the surface.

Far to the north, in Humboldt County, California, during 1892 interest revived in the oil areas which had produced a modest amount of oil in 1867. A well was drilled on J. C. Briceland's property, but was capped when it came in as a gasser. Another was drilled at the site of the first California well in 1861, near the mouth of Davis Creek. With the hole drilled, a dynamic charge was exploded at the bottom and five barrels were recovered, but for some unrevealed reason the pipe was plugged with a sapling and abandoned. Later, Union Oil leased other Humboldt County lands through a subsidiary, the Yosemite Oil Company, but the results were never disclosed.

The financial panic of 1893 caused many conservative oilmen to tend their proven wells and forget exploration temporarily. Tunnels were dug to produce small amounts of oil, but these were always minor producers and unaccountably dangerous. Several men were killed in explosions within these adits, including, in 1890, one of the Hardison brothers.

Edward L. Doheny was not a man to flinch merely because times were bad, and because of that trait he was destined to become one of the richest and most important men in the oil industry. He inspected the "brea" tar pits within the city of Los Angeles (in which the bones of prehistoric mammoths are still being found). Later, Doheny dug some small pits not far away and recovered a modest amount of oil. With his judgment sustained and faith rewarded, Doheny drilled wells with good returns. Naturally his methods were copied by hundreds of others, who drilled on every available open site.

"Wells were thick as holes in a pepper box—a well to a lot, and there were several hundred lots," Youle later recalled.

While the oil was heavy, and usually fit only for fuel, everyone drilled a well. The drilling and mess created by these wells became such a public menace that the Los Angeles city fathers forbade further oil-well drilling. It was about that time that many people discovered they needed more water than the city mains provided, so they put down "water wells." Who could help it if oil came up instead of water?

Large markets for crude oil appeared when the cities of Los Angeles and Santa Paula began using crude oil to fire the boilers in municipal power plants. But the greatest surge came when Union Oil's E. A. Edwards designed an oil burner suitable for railroad locomotives. It was tested by both the SP and the Santa Fe railroads and adopted.

Doheny brought in the Brea Canyon region of Orange County in southern California. R. C. Baker, a former associate of Doheny, and Irving Carl opened other fields in the area. The Kern County, Coalinga and the Home Oil Company's "Blue Goose" wells spouted in. And in 1900, just a breath before Spindletop showed the way, California's total annual oil production was 4,325,000 barrels.

In a history of oil discovery issued by the American Petroleum Institute, the statement is made: ". . . Never before had such fabulous figures been dreamed of in California oil circles. And a whole new century stretched invitingly just ahead. . . ."

15

The Pace Quickens

It was a curious coincidence that California oil production soared just about the same time that Spindletop blew in, and that both the Texas and California fields were discovered when the age of machinery was blossoming.

The Kern County and the Coalinga fields revived interest in other sections of California, even though the Mattole River area still refused to give oil in commercial quantities. Union Oil, with the guidance of Lyman Stewart, was developing the Santa Maria Field in southern California.

Stewart was an anachronism in the oilfields. For years he insisted that work cease on Sundays, though for technical reasons this was an expensive layoff. Stewart ambled about the fields, usually carrying a Bible, talking calmly with men whose language was punctuated with phrases which would blister a fish-tail bit. Stewart had a chapel built in the oilfields near the Union Oil wells, and was able to hire an itinerant preacher to conduct services.

Stewart's son, Bill, was something else. Young Stewart was an oilman first, last and always, liking his work and liking the men who worked with him. His father once insisted that a certain driller be fired, noting: "Any man who starts work sober and ends up the day half-drunk should be immediately discharged."

But Bill refused to fire the driller, though he didn't explain why to his father. Later, Bill told a friend: "Whoever reported his coming to work sober and getting drunk on the job is a liar. He always came to work half-drunk, then got worse. But he is a good driller, drunk or sober. That's enough."

The Los Angeles explorations, usually called the Los Angeles/Salt Lake fields because so many men from Utah owned interests, were located north of Hancock Park, an area which contained the world-famous La Brea Tar Pits.

By 1903, California was spouting a river of oil—about 25 million barrels, or more than six times the amount of oil taken in 1900—when the first burst of energy started. California issued one-quarter of the 100 million barrels produced in America during 1903, and 18 million barrels came from the Kern County wells.

In the following year, Union Oil brought in the Hartnell #1 near Santa Maria for 12,000 barrels. It produced 3 million barrels before it was necessary to apply a pump.

Though California produced many bountiful wells, the Lakeview #1, just north of Maricopa, was unique. It was a gusher of the first order.

"This well went wild . . . 9 million barrels in 18 months," W. W. Orcutt said. "On account of the heavy sand content coming up with the oil, the casing was finally worn through, and suddenly the hole caved in and the greatest gusher California ever produced was dead. Though the well was redrilled, it was never brought back. . . ."

This memorable well was drilled by "Dry Hole Charlie," whose notable lack of success accounted for his colorful nickname. But when the tremendous gusher blew in, he yelled: "My God, we've cut an artery down there."

Ed Doheny was working with Charley Canfield, another rainbow chaser who was never happy unless he was searching for something—gold, silver or oil. Doheny and Canfield had brought in several good wells, but with fluctuating markets, the collapse of some of the wells and other imponderables, they always needed to be on the move for new properties.

Doheny was fascinated with a report issued by the California State Geological Survey, written by W. A. Goodyear. All of the state's known oil seepages or springs were listed and described.

"It can scarcely be credited that he knew the full significance of the information which his annual report conveyed to the experienced oil prospector," Doheny observed. "That report was really my best guide in the later discovery of the various oil districts which it was my good fortune to open."

Doheny was extremely modest about his accomplishments. It was Doheny who made the initial discoveries in Brea Canyon, not far from Anaheim. From Brea alone, 300 million barrels of crude were recovered. The combined discoveries of Doheny and Canfield made California, at one time, America's largest producer of oil!

Because of the barrier of the Rocky Mountains, commerce in oil was more or less confined to the coastal areas of the nation. But Doheny and Canfield, like Stewart, were canny about developing markets. The Santa Fe Railroad was convinced that oil was the most efficient and economical fuel for their engines and signed a contract with Doheny and Canfield *to buy all the oil they could find* at $1 per barrel. Santa Fe officials had no idea of the finding potential of these two men. It wasn't long before the Santa Fe believed they would founder in a sea of oil found by Doheny and Canfield. Doheny was asked to cancel the contract, allowing the Santa Fe to buy some of their wells and some of the untested lease lands, all at a handsome profit. The partners later did some successful exploration for Santa Fe in Mexico.

The oil in California and all over America produced something more than mere oil: men—men of faith, men of foresight and men of ingenuity.

They worked wonders with what they found. America and her allies floated to victory in both the First and Second World Wars on oil. Airplanes, tanks, ships and guns were made mobile and workable by the oil these men brought to the surface. Even when the First World War was ended and the demand slumped, the search for oil never ceased. Mobil, Associated, Getty, Texaco, Union, Standard and most of the other major and minor companies examined every inch of California and continued to expand western oil production.

There seemed to be no limit to the oil which lay beneath the California landscape. Faith was once again affirmed on June 23, 1921, when a gusher blew in on the summit of Signal Hill in southern California. This was Alamitos #1, which was still producing a few barrels a day forty years later. It was the start of the Long Beach field, an enormous, unexpected reserve. It is estimated to have tapped the richest oil field per acre in all American history.

Signal Hill, a 365-foot-high knob of shale, was once part of the immense Los Cerritos and Los Alamitos land grants given to Manuel Nieto in 1780 by King Carlos III of Spain. For years the eminence was known as El Cerrito (Little Hill), but as it was used as an observation and signaling point for priests, pirates and smugglers, it was referred to as Signal Hill. In 1889, John Rockwell used the site as a base marker for surveying, and Coast Guard maps thereafter used the designation Signal Hill.

It was a cruel irony that one of the later owners, Abel Stearns, lost the billion-dollar acreage for $154 in back taxes. And John Temple sold his 28,000-acre Ranch Los Cerritos for 75¢ an acre following the drought of 1864.

In time, the land booms and growth of cities saw the area divided into Signal Hill view lots, which were included within the City of Long Beach. There were 55,000 people living in the area by 1920.

Oilmen skilled in geology were intrigued in Signal Hill, believing that its shape might indicate a subsurface anticline, one of the classic oil traps. Union Oil drilled to 3,449 feet in 1916 and raised nothing but dust. Four years later, Standard Oil brought in successful wells at Huntington Beach, a few miles to the south. Its Bolsa Chica Well, drilled on the rim of the Gospel Swamps, blew in with 20,000 barrels a day. It produced an estimated 7 million barrels.

Geologists studied their survey maps, noting the series of earthern humps which looked like a giant's backbone from Newport Beach all the way to Beverly Hills, a suburb of Los Angeles.

For once, Standard Oil was remiss in not following up its successes. Standard Oil geologists in 1919 reported that the region suggested a series of folding, anticlinal or dome structures, all of which might contain oil. But Standard Oil had a rigid policy of never drilling on town lots and refused to make a test well on Signal Hill, which was then entirely subdivided into town lots.

During that time, the federal government had a policy of "open door for oil"—that is, foreign countries would be allowed to explore for oil on American soil. The Royal Dutch Shell, a company largely controlled by Dutch and English investors, was therefore able to bring in the Alamitos #1.

When Shell leased 240 acres in the area of Hill and Temple streets, one cynical oil man boasted he would drink all of the oil brought from Signal Hill. Considering the millions of barrels which are still being extracted, that drink would have been a long, large one. The Shell well was brought in at 9:30 A.M. on June 23, 1921, with a 2,100-barrel gusher.

The Shell Company didn't consolidate its advantage. The company's main offices were overseas, and local operations were handled by men of limited authority. Shell was authorized to obtain leases for no more than 12 per cent, and because many Signal Hill agreements went for more, Shell wasn't able to compete successfully.

Lease purchasers and salesmen swarmed over Signal Hill making deals. Before long there were many derricks, their legs interlaced. Drills drifted into other leases, and tools were snagged in the often frantic operations.

Hourly, a Pierce-Arrow bus service brought the verdant investors to Signal Hill, where gaily colored tents served as customer booths. Tables were laden with fried chicken, salads, cakes and much strong drink. Even as the customers gulped down viands, a spieler instructed them on how money could be made without knowing anything about the oil business, or knowing much else except how to raise money to buy a lease.

Some of the people who were fortunate enough to own property on Signal Hill didn't know much about the oil business either. One man refused a lease offer of one-tenth royalty, saying he would hold out until he got one-twentieth. A barber, whose lot included a gusher, closed his tonsorial parlor and hung a sign in the window: "Cadillac salesmen, please call at my home." Cadillacs and oil were still handmaidens.

Sam Mosher was a farmer, working seventeen heavily mortgaged acres of avocado and lemon land at nearby Pico Rivera.

He was not among the throngs which flocked to see the gusher on Signal Hill. He was too busy with his own affairs, trying to scratch out a living for his family. It wasn't until mid-October 1921, several months after the well came in, that Mosher climbed into his noisy, red Buick convertible and headed for the Hill. It was more or less simple curiosity which prompted the red-haired, rugged Mosher, who held an agricultural degree from the University of California, to inspect the wells. Mosher had no idea what vast changes would be wrought once he had been smeared with the oil infection. A friend, Bob Bering, a petroleum geologist, was high on the future of Signal Hill, and was already dabbling in leases. From him Mosher learned much about the oil industry.

Both men became interested in the potential of "wet gas" which whooshed up with the gushing oil. Usually, it was allowed to escape into the air or was burned off in a tall flare stack. If processed, this gas produced "casinghead gasoline," which would bring in 23¢ a gallon in a steady market. Mosher and Bering inspected tea-kettle-sized refineries at Whittier, La Habra and Santa Fe Springs, which secured wet gas from nearby wells to obtain the gasoline.

Gradually, Mosher fell under the spell of the thundering engines, the wheeze of the boilers, the devil's tattoo beaten on the forges as dressers sharpened bits—noises which were all overridden by the ethereal calliope music of escaping gas. The more he heard it, the more he liked it.

Mosher, Bob and Lew Bering formed a partnership with Sam to arrange financing. The Berings would build the processing plant. Mosher had seen plans issued by the U.S. Bureau of Mines detailing a simple gasoline extraction plant for $4,000. Mosher also noted that the expected casinghead gasoline, when blended with other commercial fuels, gave the liquid quick-starting elements particularly vital in cold climates. Other types of gasoline, distilled directly from crude oils, didn't have the energy of the casinghead gasoline.

Mosher expected his fruit crop to supply needed funds, but a killing frost ended that hope. He borrowed the money from his father, who was highly skeptical of the plan. He pointed out that the major companies would build their plants to get the gasoline if Sam's was successful and he would be without a source of wet gas. But young Sam was persistent, and the Signal Oil & Gas Company was born, a company which was later referred to as the "Little Giant of Signal Hill," because of its dynamic policies.

With the plant beginning to become a reality, Mosher began to contract with producers in the area. He offered one-third royalties on all gasoline sales made by the Signal Oil Company and free residue gas to be used as boiler fuel in exchange for tying their gathering lines to wells issuing wet gas.

Mosher's proposals met with cold indifference. There was general lack of trust in the inexperienced boys, lack of willingness to commit themselves to a life-of the-field contract with a nonexistent plant. Finally the San Martinez Oil Company agreed to let Mosher lay gathering lines and collect the natural gas.

Mosher was excited. Henry Ford's Model T was selling for less than $300 in 1922, putting it well within the reach of the public. The demand for these cars was strong and getting stronger.

The Bendix self-starters were not generally available, and this was the day of the hand cranks and fractured forearms. Windshield wipers were hand-operated, rear-view mirrors were a curiosity and not all tires came with demountable rims. Filling stations—service station was a name five years in the future

—then consisted of five gallon hand-cranked pumps where free tire air and comfort stations were also available.

But Mosher foresaw that America would be criss-crossed with highways and these would be used by cars of many and varied makes. All, hopefully, would be using Mosher's gasoline in prodigious quantities.

Mosher went doggedly ahead, dreaming his dreams, buying used equipment in an effort to conserve their dwindling capital. Despite all obstacles, the plant was assembled sixty days after starting.

Bering hadn't come up with any contracts, so Mosher went directly to Shell officials, asking them to inspect his plant and allow him to use the waste gas from Shell wells.

"I can knock more gasoline out of the wet gas with a bent stick than you can extract with that tea-kettle outfit," Shell engineers scoffed. But they finally allowed Mosher to have their wet gas until such time as Shell constructed their own plant. It was a major victory and assured the success of Signal Oil.

By May, 1922, Mosher's plant was producing a thin trickle of water-clear gasoline, about 250 gallons a day. Even at 23¢ a gallon this amount wasn't meeting daily expenses. There was only one solution: expand.

Additional stock was issued for $100,000 in $1 shares. Forty thousand shares were offered to the public, and they sold briskly. The plant was expanded. But there was still trouble. Sales of the finished product didn't always match production, and the remaining gasoline was a wildly volatile product with a high evaporation ratio.

Mosher later described it as being like selling steam from an open bucket.

To correct the imbalance, raw gasoline was covertly dumped into ditches running parallel with the city streets. If they had ever touched off, Signal Hill would have become an inferno which would have burned for weeks.

But the gasoline market finally caught up with and exceeded production, and with it the "Little Giant of Signal Hill" was certain of success. It was another example of men with faith and courage being willing to back their judgment with everything they had.

California oil production continued in an upward spiral from the advent of the Signal Hill gusher, which eventually supported more than 1,200 wells.

Oilmen are as restless as any other dreamers, and the tidelands tempted them. Many years before (offshore drilling began in 1894), oil had been obtained there. Perhaps there was more. With improved methods of drilling, the whipstock, the ability to drill on a slant, ensured the recovery of oil in previously inaccessible areas. (Years later, floating rigs would be developed and used successfully.) Immediately there was a public outcry about the prospects of damage to fish and other undersea life, the beaches or other property. There was an immediate legal contention among the federal, state and local governments as to who held final jurisdiction and in what areas, especially in view of the immense oil royalties which were expected to ensue. In California, the Attorney General was given temporary authority, but that was later usurped through extended litigation between the political units. Almost from the time the first drill probed underwater, legal conservation, oil industry and other contentions have raged. No end is actually in sight as the political fortunes ebb and flow through the years.

California was being criss-crossed with a lattice-work of pipelines to accommodate the production from remote areas and gather the oil for urban centers, where most of the refineries were located. Diesel engine power was used to push the oil through long, uphill pipelines, and the first submarine lines to load tankers at sea were put down by the Associated Oil Company—one from Ventura, another from Monterey, a small port south of San Francisco.

In 1953, California first achieved its million-barrels-a-day production for one entire year, coming in with 365,085,000 barrels. Even with that enormous production of oil, California was gradually becoming an importer of both oil and natural gas because of the population explosion and industrial expansion.

It is curious that California would now be considered relatively oil-poor, with 9,536 miles of pipeline (in 1968), thirty-two operating refineries and at least $7.5 billion dollars invested in the California oil industry.

16

Suitcase Rock

When formations are reached which indicate there is no further hope of finding oil, the strata is called "suitcase rock." This is the time to pack up and depart for more productive fields, and it happens frequently, since no more than one in nine wells drilled ever produces oil in commercial quantities.

There are estimates of at least a 400-billion-barrel ultimate reserve underground, and discoveries are being made each day by the major companies and the individual wildcatter. Despite this, apprehensions are frequently voiced that the world will someday be exhausted of all its oil.

More uses are found for petroleum products almost daily, and the search to fill that demand goes on. Alaska and other remote American areas are being prospected, and the most promising areas have been uncovered by joint operations of the Atlantic-Richfield and Humble Oil companies. Discussing the successful well at Prudhoe Bay, Alaska, officials called it "one of the largest petroleum accumulations known to the world today."

Finds in such remote areas have to be extensive to justify the expense, hardships and difficulties of extracting oil and getting it to market within competitive price ranges. Oilmen have always been cautious about drilling in Alaska because of the prohibitive costs, infrequent successes and the need for a major discovery in order to have commercial significance. The first hole drilled was the Susie #1, on a 90,000-acre lease at a cost of $4 million. It was abandoned at 13,500 feet, a duster. The "discovery well," the Prudhoe Bay #1, cost $5 million.

Despite the added hardships, these wildcat, exploratory wells can only be drilled in the winter. Warmer weather turns the tundra into a quagmire, and the heavy drilling rigs sink out of sight as the surface thaws. Even the drilling mud has to be cooled before being recirculated to avoid thawing the permafrost surface surrounding the drilling hole. Thawed, the drilling rig might drop into the hole.

Because of the winter temperatures, tools break like twigs, and most engines have to be kept running twenty-four hours a day to prevent freezing. Even so, they have to be lubricated with special oils.

Drilling rigs rest on pilings around which water has been poured. The water freezes immediately, giving the structure a firm footing. One of the most difficult problems is the delivery of crude oil after it has been brought to the surface. Any oil freezes solid at minus 65 degrees. Ships can reach these remote areas only occasionally. Air transport is too expensive, and railroads, which many favor, would be extremely expensive to build, and even then unable to reach some areas. The solution seems to be in an underground pipeline in the permafrost zone, which remains close to a constant 20 degrees above zero. Oil from this area has a pour-point about 30 degrees colder than that, assuring that the fluid would move along the lines unless there was too much paraffin in the oil, which would tend to consolidate and clog the lines, much like the hardening of human arteries.

But oilmen, since Colonel Drake, have always overcome their problems.

Off the Long Beach, California, shoreline, five major oil

companies—Texaco, Humble, Union, Mobil and Shell, (a grouping usually called "THUMS")—have cooperated in drilling 928 wells on a 6,480-acre offshore lease. By the year 2000, the companies expect to have spent $1 billion and recovered an oil reserve of 1.2 billion barrels of crude oil.

But those astronomical figures are not what makes the project unique. The wells will be drilled on four man-made islands, named after deceased astronauts Grissom, White, Chaffee and Freeman. The islands have been landscaped with palm trees and other plants. The derricks are enclosed in what appear to be high-rise apartments. They conceal all of the drilling operations, which will send some of the bits on a 75-degree slant to reach the outer edges of oil sands. All of the massive drilling equipment is mounted on movable structures. When a well is brought in, the "high-rise buildings" are shifted to a new site with never a derrick or pump being seen from shore. Contracts require that all pumping equipment be placed below ground level so that the islands continue to present an attractive appearance from the shore. It is no doubt the most ingenious and cosmetic oil recovery plan ever devised.

There is need for perpetual search. There are about 3,000 separate products extracted from crude oil, natural gas and natural gas liquids—the raw materials considered "petroleum." With the application of various skills these can become a red dress, stove fuel, fertilizer pellets or alcohol.

Isopropanol, an alcohol, was the first petrochemical—developed during 1920. Shortly thereafter, antifreeze for cars was developed; nylon came along in 1937.

The reason so many products can be developed from this substance is that petroleum is composed essentially of hydrogen and carbon atoms. These can be arranged by petrochemists in an almost endless variety of molecules, which are simply combinations of atoms. Petroleum also has a small amount of sulfur and nitrogen, both of which amplify the uses to which petroleum can be put. An estimated $500 million dollars is spent each year for petroleum research.

Sudden need is often the reason for the vast expansion of petroleum products. When World War II erupted, the supplies of raw rubber from Asia were cut off by the Japanese. And most of the world rolls on rubber. A synthetic, butadiene, was developed by the Phillips Petroleum Company. Many other companies had important parts in the development and later production of synthetic rubber. The companies worked together in providing the military forces with huge amounts of carbon black, toluene and ammonia—all derivatives of petroleum—for explosives.

The desperate need for these items was brought about by the war, and the petrochemical industry responded during 1940 with a multitude of new refining processes. Plants were designed to discover and produce new products, an extremely expensive process unless a firm market can be anticipated to justify the expenses of each company.

Among these new products were the powerful fuels to feed into military aircraft, which required more sophisticated refining methods. Techniques used to develop these special fuels usually rearrange the structure of the hydrocarbons. These processes include catalytic cracking, polymerization, hydrogenation and alkylation, some of which were pioneered by the Phillips Petroleum Company.

It was the increased production of high-octane gasoline which showed the way for a number of the petrochemicals. Because of the expanding need for these items, the expense of new plants, more researchers and engineers is always justified.

When the really great surge in the growth of the petrochemical industry began in about 1945, about 3.3 billion pounds of petrochemical products was produced, worth an estimated $211 million dollars. These figures expanded manyfold within a few years and are still growing. There is no end in sight.

Such explosive growth was brought about through three main factors. Many of the petrochemicals have been made to replace more expensive metal, wood, glass and textiles. Such chemical wizardry offsets the chance that the United States might ever be denied any of these vital items, because of war or for any other reason. Such synthetics are usually cheaper, and even better, than the original products. Synthetic rubber is a prime example of this.

The importance of these petrochemicals was emphasized by the development of new substances—the derivatives which have been discovered, such as plastic, another petrochemical end product. The first plastic product was, incidentally, a billiard ball, made in 1868.

With the discovery of these new substances, combined with a growing acceptance of them, a strong competitive situation has arisen as well. With the race on for the best product at the best price, more efficient refining, research and marketing procedures have also expanded, with concomitant great benefit to the public.

Use of wet natural gas has previously been mentioned in connection with the Signal Oil Company and others. Though the breakthrough was slow because of certain technical chemical problems, the present uses of natural gas extractions and synthetics are almost without number.

One of the first of consequence to the public was a stable antifreeze solution for cars. Up to 1926, wood alcohol had been used in radiators. The fluid evaporated rapidly, making replacement necessary at added costs. Another type of alcohol was extracted from natural gas, and antirust and antileak ingredients were added along with a noncorrosive substance; the product was offered under the name of Prestone, still one of the leaders in this field. Once again the public had benefited from the research being done by the oil industry.

Natural gases give up a myriad of substances, either directly or after repeated fractionalization and combination with other materials. Solvents of all sorts, such as acetone (used in fingernail polish remover and many other products), ethers for anesthetics, glycerine for soaps, numerous alcohols with specific uses, the aldehydes (including formaldehyde, used in embalming), are but a few of the better-known items. The processor of natural gas works closely with the petroleum industry, in that each produces some compounds valuable to the other—directly or in linking certain chemical processes which would be impossible without the natural gas fractions of synthesis.

The development of these petrochemicals is expensive and extremely intricate. The chemical methods used constitute a complete study in themselves, having no detailed place in the discovery of oil. But suffice it to say that unimagined comforts, medicines and other benefits to mankind are already being studied in the flasks and test tubes of the petrochemists, whose imaginations soar beyond vision.

The extraction of these petrochemicals requires refining and continuous processing in various plants. It is because of this that almost all petrochemicals are produced along what is called "the Golden Crescent," a strip of coastal land between New Orleans, Louisiana, and Brownsville, Texas. A large amount of known gas and oil reserves is located in the Crescent. The Crescent is also immediately accessible to ships which touch every port in the world. There are ample supplies of fresh water, vital to the refining processes. The flat topography of the area lends itself to the use of a maze of pipelines which connect many of the plants. There are so many pipelines between Houston and Beaumont that the area is usually called "the spaghetti bowl." Most of the major companies are represented in the Golden Crescent area.

Because of the vast scope of petrochemical products, it is difficult to name the most important.

Ammonia is the largest-volume inorganic product. It is the source of nitrogen for soil treatment, and is also used in explosives, cleaning fluids, yeast nutrients and refrigerants.

Plastics and the resin groups are of great importance among the organic petrochemical products. Many of these, like ammonia, have helped the farmers. Farm produce is often packaged in plastic sheaths. Irrigation water is frequently piped through plastic tubes. Some flavoring agents, preservatives and adhesives are of petrochemical origin.

One of the major synthetics is rubber, which is now considered of better quality and durability than the natural product from rubber trees. This rubber is also used for such things as coating wires and cables, shoe soles, flooring and an innumerable series of other items which are used every day.

A second is carbon black, which is used in the synthetic rubber tires to give them color, long life and toughness. Ink, paint, typewriter ribbons, records and pencils are just a few of the uses for carbon black.

Synthetic fibers, of which nylon is the best known, are vital to modern living. Umbrellas, rope, diapers, draperies and rugs are a few substances which are made from the synthetic fibers.

While these cover a large spectrum of petrochemicals, consider the following, all a part of everyday life: lipstick, margarine, shampoo, detergent soap, lotions, food preservatives, paste, paint, insect sprays, fruit flavorings, aerosol bombs, weed killers and fire extinguishers.

The vistas of science have been widened because of the petrochemicals. Lives are extended through a number of drugs compounded from petroleum to treat illness or combat infections. Plastic casts can now be used instead of the old, cumbersome plaster casts to set broken bones.

Transportation of various materials is made easier and cheaper through the use of plastic containers. Many parts are made of plastic, which is usually lighter than most metals and often as strong.

Some solid rocket fuels are composed in part of rubbery gels of petroleum-derived chemicals. Missiles themselves use numerous plastic parts, and the astronauts wear protective garments which are fabricated from synthetic fibers.

With the exploration of space and the seas expanding at a furious rate, the use of petrochemicals will be in heavy demand for the foreseeable future.

Though it may occur in certain individual instances, there is actually no "suitcase rock" ahead for the oil industry. The tocsin will always be: "Dad's struck ile."

Glossary

Anticline: a fold in rocks which makes the formation look like an inverted bowl. Oil is often trapped within limestones or sandstones.

Asphalt: a chemical combination of hydrocarbons derived from the refining petroleum. Although the bitumens which compose asphalt may be in liquid or solid state, asphalt is defined as a semisolid.

Cable-tool drilling: a percussion type of well drilling. Bits are raised and lowered by various devices, and the impact gradually grinds a hole. This method has been largely replaced by rotary drilling rigs.

Carbon black: a deposit of carbon resulting when oil or gas is burned in an insufficient supply of air.

Casing: a pipe inserted into the well to prevent caving or the intrusion of water or other substances. These casings are cemented in place to ensure permanency.

Coal oil: an archaic term for kerosine. It now primarily means oil obtained by the destructive distillation of bituminous coal.

Cores: rock samples brought to the surface by special bits. Geologists study these for clues to oil.

Crown block: where pulleys are installed at the top of the derrick to lift tools, pipe and other things.

Derrick: a lattice-work steel tower which sits astride the well as drilling operations are conducted, and which is later dismantled if economically feasible. They are tall enough to handle several sections of pipe being withdrawn from the well and are generally more than one hundred feet high (though this figure is not fixed by anything other than conditions at hand). The four-sided structure is more than twenty feet square at the base and tapers to about five feet at the top.

Doodlebug: a pseudoscientific method of locating oil. Use of these "divining rods" or "witching sticks" has been largely discredited, and the term is now one of mild derision.

Duster, or dry hole: a well which produces neither gas nor oil.

Fault: a rock slippage which fractures a rock stratum and occasionally forms a trap to accommodate oil, especially in sandstone or limestone formations.

Fishtail bits: a drilling tool which looks like a fishtail, with a "V" indentation in it, and works like an augur to bore through the softer formations, such as sand, gravel or clay.

Gusher: any well which, because of the large natural flow, does not require pumping.

Magnetometer: a delicate instrument which measures the force and direction of the earth's gravitational pull, measurements which provide the petroleum engineers with clues about oil locations.

Methane: the principal ingredient in natural gas. The result of organic decay, it is also called "marsh gas."

Oil sand: sandstone stratum which contains oil.

Paraffin: a substance found in certain oils and tars which may be extracted by distillation. Such oils are said to be "paraffin-based," in contrast with "asphalt-based" oils.

Perforation: blowing holes in the casing to let oil seep in.

It is believed that the Roberts Torpedo was the first successful use of explosives to make these perforations.

Pool: a loose, discredited term to describe reservoirs of underground oil. Oil is not found in pools but in oil sands and certain types of rocks.

Prospecting: a term applied to anyone searching for oil, just as it referred to the rainbow chasers in the Gold Rush days. Before 1900, oil was sought more by guesswork than by geology, but now prospecting is an accredited science which uses all of the most modern scientific instruments. Even so, a large number of dusters are drilled. Only one in nine wells produces oil in commercial quantities.

Rotary drilling: a method which has largely replaced the cable-tool method. Bits are attached to the end of the drill pipe, and the entire pipe is rotated by an engine on the platform floor. There are dozens of different bits, which are used in various strata or situations.

Royalty payments: money paid to the landowner or royalty owners when a successful well is brought in. A fixed percentage is agreed upon before the well is drilled.

Salt dome: a giant accumulation of salt which has pushed upward, frequently forming oil traps along its sides or top. Domes are found in large quantities along the Golden Crescent (the Texas-Louisiana coast).

Seismograph: an instrument which measures the shock waves forced through rock strata by the detonation of explosive charges strategically placed by geologists. Electronic waves are measured in offshore exploration to protect undersea life.

Shooting the well: the practice of lowering high-level explosives into a well and exploding them at oil-sand levels in the hopes of improving the oil yields. This work is done by "shooters" and is very hazardous. Explosives are also used to extinguish oil-well fires.

Spudding in: the initial operations necessary to begin a well; the foundation. The first few feet of drilling is done with a "spudding bit."

Suitcase rock: strata in a producing area which indicate that no further oil is to be found.

Tank farm: a location on which numerous oil storage tanks are situated.

Tool dresser: the driller's helper, who dresses (reshapes or sharpens) bits.

Tool pusher: usually the oilfield foreman in charge of rig building and drilling. He may be in charge of more than one crew.

Torsion balance: another delicate instrument designed to aid in the accurate location of oil. The gravitational stresses at various places are measured.

Walking beam: a beam set on a fulcrum in such a way that it can be moved up and down though it is attached at both ends. It is used largely in cable-tool drilling. The same term is used to describe a similar teetering steel beam in the pumping of established wells.

Wet gas: the natural gas from which gasoline is easily extracted. A "dry gas" is one without gasoline or in such a mixture that it is expensive and difficult to separate.

Wildcatter: a card-carrying optimist who seeks oil in previously unexplored areas or in other unproven terrain.

Selected Bibliography

Asbury, Herbert. *The Golden Flood, An Informal History of America's First Oil Field*. New York: Knopf, 1941.

Boatright, Mody C. *Folklore of the Oil Industry*. Dallas, Tex.: Southern Methodist University Press, 1963.

Clark, James, and Halbouty, Michael. *Spindletop*. New York: Random House, 1952.

Dolson, Hildegarde. *The Great Oildorado*. New York: Random House, 1959.

Duncan, Bob. *The Dicky Bird Was Singing*. New York: Rinehart, 1952.

Glasscock, Carl B. *Then Came Oil*. New York: Bobbs-Merrill, 1938.

James, Marquis. *The Texaco Story, 1902/1952*. Privately Printed, 1953.

Knowles, Ruth Sheldon. *The Greatest Gamblers*. New York: McGraw-Hill, 1959.

Miller, Max. *Speak to the Earth*. New York: Appleton-Century, 1955.

Rister, Carl C. *Oil! Titan of the Southwest*. Norman, Okla.: University of Oklahoma Press, 1949.

Tait, Samuel, Jr. *The Wildcatters*. Princeton, N.J.: Princeton University Press, 1946.

Tompkins, Walker. *Little Giant of Signal Hill*. Englewood Cliffs, N.J.: Prentice-Hall, 1964.

Welty, Earl, and Taylor, Frank J. *The 76 Bonanza.* Menlo Park, Calif.: Lane Book Company, 1966.

Williamson, Harold F., Andreano, Ralph, Daum Arnold R., and Klose, Gilbert C. *The American Petroleum Industry, 1899/1959.* Evanston, Ill.: Northwestern University Press, 1963.

Index

About the Authors

The collaboration of Bob and Jan Young began at the University of California at Los Angeles, where they met as undergraduates and found they shared a common interest in writing. Following their marriage in 1940, they spent most of the next ten years in the newspaper field. In 1950 they turned to free-lance writing, concentrating first on the magazine field. In 1958 they published their first two books for children and now, after numerous other books, consider writing for young people their major interest.

Both are native Californians. Bob was born November 6, 1916 in Chico; attended Sacramento schools, UCLA and graduated from the University of Nevada. Jan was born March 6, 1919 in Lancaster; attended Pt. Loma and South Pasadena public schools and UCLA. They are the parents of four children.

Both are enthusiastic fishermen and they now make their legal residence in Ferndale on the redwood coast of northern California. Winters, they continue to spend in their former family home in Whittier in southern California. Other hobbies, when they find time from their writing, include sculpturing, painting and traveling throughout the West to the many historic spots that play such an important part in their writing.

PRESCOTT HIGH SCHOOL
3 2793 00005 5766